HVAC
Engineer's
Handbook

HVAC Engineer's Handbook

Eleventh edition

F. Porges
LL.B, BSc(Eng), CEng, FIMechE, MIEE, FCIBSE

OXFORD   AUCKLAND   BOSTON   JOHANNESBURG   MELBOURNE   NEW DELHI

Butterworth-Heinemann
Linacre House, Jordan Hill, Oxford OX2 8DP
225 Wildwood Avenue, Woburn, MA 01801-2041
A division of Reed Educational and Professional Publishing Ltd

A member of the Reed Elsevier plc group

First published as *Handbook of Heating, Ventilating and Air Conditioning* 1942
Second edition 1946
Third edition 1952
Fourth edition 1960
Fifth edition 1964
Sixth edition 1971
Seventh edition 1976
Eighth edition 1982
Ninth edition 1991
Tenth edition 1995
Reprinted 1997, 1998
Eleventh edition 2001
Transferred to digital printing 2004
© F. Porges 1982, 1991, 1995, 2001

British Library Cataloguing in publication Data
A catalogue record for this book is available from the British Library

Library of Congress Cataloguing in publication Data
A catalogue record for this book is available from the Library of Congress

ISBN 0 7506 4606 3

Contents

Preface

This book contains in a readily available form the data, charts and tables which are regularly required by heating, ventilating and air conditioning engineers in their daily work.

The data is presented in a concise manner to enable it to be applied directly in the actual daily work of the HVAC engineer. The book is designed for everyday use and a comprehensive bibliography has been included for the benefit of those who wish to pursue the theoretical side of any particular topic.

For this edition some errors have been corrected, the explanatory notes on the psychrometric chart have been improved and the chart in previous editions has been replaced, with permission, by the well known CIBSE chart. Additional data has been included on design temperatures and ventilation rates and information has been inserted on precautions against legionellosis in both hot water systems and air conditioning plant. The data on duct thicknesses and sizes has been revised to conform to current practice. A new section has been included on natural ventilation and the information on types of refrigeration compressors has been expanded. The data on refrigerants has been completely revised to list the new non-CFC and non-HCFC refrigerants. Practising engineers will still meet old plant which contains refrigerants which are now obsolete or obsolescent, and therefore the properties of the more important of these are also given.

The policy of previous editions of giving tabulated data in both SI and Imperial units has been continued although theoretical expressions are generally given only in SI units.

F. Porges

The author would like to acknowledge the help of Mrs Christine Tenby in the compilation of the index.

1 Abbreviations, symbols and conversions

Symbols for units

m	metre	s	second	st	stoke
mm	millimetre	min	minute	J	joule
μm	micrometre	h	hour	kWh	kilowatt hour
	(formerly	d	day	cal	calorie
	micron)	yr	year	Btu	British
in	inch	kg	kilogram		thermal unit
ft	foot	t	tonne	W	watt
yd	yard	lb	pound	V	volt
m^2	square metre	gr	grain	A	ampere
mm^2	square millimetre	cwt	hundred weight	VA	volt ampere
a	acre	N	newton	K	kelvin
ha	hectare	kgf	kilogram force	°C	degree Celsius
in^2	square inch	pdl	poundal	°F	degree Fahrenheit
ft^2	square foot	lbf	pound force	°R	degree Rankine
m^3	cubic metre	Pa	pascal	dB	decibel
l	litre	m^2/s	metre		
in^3	cubic inch		squared per		
ft^3	cubic foot		second		
gal	gallon				

Symbols for physical quantities

l	length	α	attenuation	T	thermo-
h	height		coefficient		dynamic
b	width	β	phase		temperature
r	radius		coefficient	θt	common
d	diameter	m	mass		temperature
AS	area	ρ	density	C_p	specific heat
V	volume	d	relative		capacity at
t	time		density		constant
T	period (time	F	force		pressure
	of one cycle)	W	weight	C_v	specific heat
uvw	velocity	M	moment		capacity at
ω	angular	h	pressure		constant
	velocity	w	work		volume
a	acceleration	p	power	U	thermal
g	acceleration	η	efficiency		trans-
	due to	ν	kinematic		mittance
	gravity		viscosity	k	thermal
					conductivity

1

Multiples and sub-multiples

$\times 10^{12}$	tera	T	$\times 10^{-1}$	deci	d
$\times 10^{9}$	giga	G	$\times 10^{-2}$	centi	c
$\times 10^{6}$	mega	M	$\times 10^{-3}$	milli	m
$\times 10^{3}$	kilo	k	$\times 10^{-6}$	micro	μ
			$\times 10^{-9}$	nano	n
			$\times 10^{-12}$	pico	p

Abbreviations used on drawings

BBOE	bottom bottom opposite ends (radiator connections)	LSV	lockshield valve
		MV	mixing valve
CF	cold feed	MW	mains water
CW	cold water	NB	nominal bore
DC	drain cock	NTS	not to scale
EC	emptying cock	PR	primary (hot water flow)
F	flow	R	return
FA	from above	SEC	secondary
TA	to above	TA	to above
FS	fire service	TB	to below
FTA	from and to above	TBOE	top bottom opposite ends (radiator connections)
FTB	from and to below		
FW	fresh water	TBSE	top bottom same end
GV	gate valve	TW	tank water
HTG	heating	TWDS	tank water down service

Standard sizes of drawing sheets

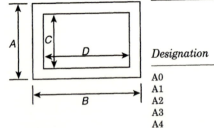

	Size of sheet		Size of frame	
	A	B	C	D
Designation	mm	mm	mm	mm
A0	841	1189	791	1139
A1	594	841	554	804
A2	420	594	380	554
A3	297	420	267	390
A4	210	297	180	267

Recommended scales for drawings

1:1	1:10	1:100	1:1000
1:2	1:20	1:200	
1:5	1:50	1:500	

Symbols on drawings (based on BS 1553)

———	PIPE		ANGLE VALVE
— — — — ·	PIPE BELOW GROUND		RELIEF VALVE
— · —	PIPE AT HIGH LEVEL		ANGLE RELIEF VALVE
////////	EXISTING PIPE TO BE REMOVED		NON-RETURN VALVE
—+—	CROSSING, UNCONNECTED		THREE-WAY VALVE
—•—•	JUNCTION, CONNECTED		FOUR-WAY VALVE
——→	INDICATION OF FLOW DIRECTION		FLOAT OPERATED IN LINE VALVE
FALL 1 : 200	INDICATION OF FALL		GLOBE VALVE
=====	HEATED OR COOLED		BALL VALVE
	JACKETED		BELLOWS
	GUIDE		STRAINER OR FILTER
	ANCHOR		TUNDISH
	IN LINE VALVE (ANY TYPE)		OPEN VENT

Symbols on drawings *(continued)*

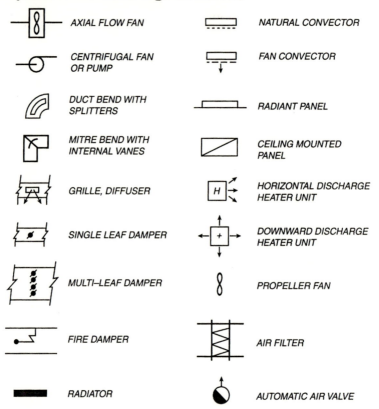

AXIAL FLOW FAN

CENTRIFUGAL FAN
OR PUMP

DUCT BEND WITH
SPLITTERS

MITRE BEND WITH
INTERNAL VANES

GRILLE, DIFFUSER

SINGLE LEAF DAMPER

MULTI–LEAF DAMPER

FIRE DAMPER

RADIATOR

NATURAL CONVECTOR

FAN CONVECTOR

RADIANT PANEL

CEILING MOUNTED
PANEL

HORIZONTAL DISCHARGE
HEATER UNIT

DOWNWARD DISCHARGE
HEATER UNIT

PROPELLER FAN

AIR FILTER

AUTOMATIC AIR VALVE

Conversions

Length

1 in = 25.4 mm
   = 0.0254 m
1 ft = 0.3048 m
1 yd = 0.9144 m
1 m = 3.2808 ft
   = 1.0936 yd
1 mm = 0.03937 in

Area

$1 \text{ in}^2 = 6.452 \text{ cm}^2$
   $= 6.452 \times 10^{-4} \text{ m}^2$
$1 \text{ ft}^2 = 0.0929 \text{ m}^2$
$1 \text{ yd}^2 = 0.836 \text{ m}^2$
$1 \text{ ac} = 4840 \text{ yd}^2$
   $= 0.4047 \text{ ha}$
$1 \text{ mm}^2 = 1.55 \times 10^{-3} \text{ in}^2$
$1 \text{ m}^2 = 10.764 \text{ ft}^2$
   $= 1.196 \text{ yd}^2$
$1 \text{ ha} = 10^4 \text{ m}^2$
   $= 2.471 \text{ ac}$

Volume

$1 \text{ in}^3 = 16.39 \text{ cm}^3$
   $= 1.639 \times 10^{-5} \text{ m}^3$
$1 \text{ ft}^3 = 0.0283 \text{ m}^3$
   $= 6.23 \text{ gal}$
$1 \text{ yd}^3 = 0.7646 \text{ m}^3$
$1 \text{ gal} = 4.546 \text{ l}$
   $= 4.546 \times 10^{-3} \text{ m}^3$
   $= 0.16 \text{ ft}^3$
$1 \text{ pint} = 0.568 \text{ l}$
$1 \text{ U.S. gal} = 0.83 \text{ Imperial gal}$
$1 \text{ cm}^3 = 0.061 \text{ in}^2$
$1 \text{ m}^3 = 35.31 \text{ ft}^3$
   $= 1.308 \text{ yd}^3$
   $= 220.0 \text{ gal}$
$1 \text{ l} = 0.220 \text{ gal}$

Mass

1 grain = 0.000143 lb
   = 0.0648 g
1 lb = 7000 grains
   = 0.4536 kg
   = 453.6 g

1 g = 15.43 grains
   = 0.0353 oz
   = 0.002205 lb
1 kg = 2.205 lb
1 tonne = 1000 kg
   = 0.984 tons

Content by weight

1 g/kg = 7.0 gr/lb
1 gr/lb = 0.143 g/kg

Density

$1 \text{ lb/ft}^3 = 16.02 \text{ kg/m}^3$
$1 \text{ kg/l} = 62.43 \text{ lb/ft}^3$
$1 \text{ kg/m}^3 = 0.0624 \text{ lb/ft}^3$

Velocity and volume flow

1 ft/min = 0.00508 m/s
1 m/s = 196.85 ft/min
1 kg/s (water) = 13.20 gal/min
$1 \text{ m}^3/\text{s} = 2118.9 \text{ ft}^3/\text{min}$
$1 \text{ ft}^3/\text{min} = 1.7 \text{ m}^3/\text{h}$
   = 0.47 l/s
1 l/s = 792 gal/h
   = 13.2 gal/min

Heat flow

1 Btu/h = 0.293 watt
1 kW = 1000 J/s
   $= 3.6 \times 10^6 \text{ J/h}$
   = 1.360 metric
   horse power
   = 737 ft lb/s
   = 3412 Btu/h
   = 860 kcal/h
$1 \text{ kcal/h} = 1.16 \times 10^{-3} \text{ kW}$
$1 \text{ Btu/ft}^2 = 2.713 \text{ kcal/m}^2$
   $= 1.136 \times 10^4 \text{ J/m}^2$
$1 \text{ Btu/ft}^2 \text{ h} = 3.155 \text{ W/m}^2$
$1 \text{ Btu/ft}^3 \text{ h} = 10.35 \text{ W/m}^3$
$1 \text{ Btu/ft}^2 \text{ °F} = 4.88 \text{ kcal/m}^2 \text{ K}$
   $= 2.043 \times 10^4 \text{ J/m}^2 \text{ K}$
$1 \text{ Btu/ft}^3 = 8.9 \text{ kcal/m}^3$
   $= 3.73 \times 10^4 \text{ J/m}^3$

Conversions *(continued)*

$$1\ \text{Btu/lb} = 0.556\ \text{kcal/kg}$$
$$= 2326\ \text{J/kg}$$
$$1\ \text{kcal/m}^2 = 0.369\ \text{Btu/ft}^2$$
$$1\ \text{kcal/m}^2\ \text{K} = 0.205\ \text{Btu/ft}^2\ {}^\circ\text{F}$$
$$1\ \text{kcal/m}^3 = 0.112\ \text{Btu/ft}^3$$
$$1\ \text{kcal/kg} = 1.800\ \text{Btu/lb}$$
$$1\ \text{ton refrigeration} = 12.000\ \text{Btu/h}$$
$$= 3.516\ \text{kw}$$
$$1\ \text{ft}^2\ \text{h}\ {}^\circ\text{F/Btu} = 0.18\ \text{m}^2\ \text{K/w}$$
$$1\ \text{ft}^2\ \text{h}\ {}^\circ\text{F/Btu in} = 6.9\ \text{m K/w}$$
$$1\ \text{Btu/h ft}^2\ {}^\circ\text{F} = 5.68\ \text{W/m}^2\ \text{K}$$

Pressure

$$1\ \text{atm} = 1.033 \times 10^4\ \text{kg/m}^2$$
$$= 1.033\ \text{kg/cm}^2$$
$$= 1.013 \times 10^2\ \text{kN/m}^2$$
$$= 1.013\ \text{bar}$$
$$= 14.7\ \text{lb/in}^2$$
$$= 407.1\ \text{in water at } 62^\circ\text{F}$$
$$= 10.33\ \text{m in water at } 62^\circ\text{F}$$
$$= 30\ \text{in mercury at } 62^\circ\text{F}$$
$$= 760\ \text{mm mercury at } 62^\circ\text{F}$$

$$1\ \text{lb/in}^2 = 6895\ \text{N/m}^2$$
$$= 6.895 \times 10^{-2}\ \text{bar}$$
$$= 27.71\ \text{in water at } 62^\circ\text{F}$$
$$= 703.1\ \text{mm water at } 62^\circ\text{F}$$
$$= 2.0416\ \text{in mercury at } 62^\circ\text{F}$$
$$= 51.8\ \text{mm mercury at } 62^\circ\text{F}$$
$$= 703.6\ \text{kg/m}^2$$
$$= 0.068\ \text{atm}$$

$$1\ \text{kg/m}^2 = 1.422 \times 10^{-3}\ \text{lb/in}^2$$
$$= 9.81\ \text{N/m}^2$$
$$= 0.0394\ \text{in water}$$
$$= 1\ \text{mm water}$$
$$= 0.0736\ \text{mm mercury}$$
$$= 0.9681 \times 10^{-4}\ \text{atm}$$

$$1\ \text{N/m}^2 = 0.1450 \times 10^{-3}\ \text{lb/in}^2$$
$$= 1 \times 10^{-5}\ \text{bar}$$
$$= 1 \times 10^{-2}\ \text{mbar}$$
$$= 4.03 \times 10^{-3}\ \text{in water}$$
$$= 0.336 \times 10^{-3}\ \text{ft water}$$
$$= 0.1024\ \text{mm water}$$

$$= 0.295 \times 10^{-3}\ \text{in mercury}$$
$$= 7.55 \times 10^{-3}\ \text{mm mercury}$$
$$= 0.1024\ \text{kg/m}^2$$
$$= 0.993 \times 10^{-5}\ \text{atm}$$

$$1\ \text{kN/m}^2 = 1 \times 10^{-2}\ \text{bar}$$

$$1\ \text{in water} = 0.0361\ \text{lb/in}^2$$
$$= 249\ \text{N/m}^2$$
$$= 25.4\ \text{kg/m}^2$$
$$= 0.0739\ \text{in mercury}$$

$$1\ \text{mm water} = 1.42 \times 10^{-3}\ \text{lb/in}^2$$
$$= 9.80\ \text{N/m}^2$$
$$= 1\ \text{kg/m}^2$$
$$= 0.0736\ \text{mm mercury}$$
$$= 0.9677 \times 10^{-4}\ \text{atm}$$

$$1\ \text{in mercury} = 0.49\ \text{lb/in}^2$$
$$= 3378\ \text{N/m}^2$$
$$= 12.8\ \text{in water}$$

$$1\ \text{mm mercury} = 0.0193\ \text{lb/in}^2$$
$$= 133\ \text{N/m}^2$$
$$= 12.8\ \text{mm water}$$

$$1\ \text{bar} = 1 \times 10^5\ \text{N/m}^2$$
$$= 14.52\ \text{lb/in}^2$$
$$= 100\ \text{kN/m}^2$$
$$= 10.4\ \text{mm w.g.}$$

$$1\ \text{Pa} = 1\ \text{N/m}^2$$

Energy and heat

$$1\ \text{joule} = 1\ \text{watt second}$$
$$= 1\ \text{Nm}$$
$$= 0.74\ \text{ft lb}$$
$$= 9.478 \times 10^{-4}\ \text{Btu}$$

$$1\ \text{Btu} = 1.055 \times 10^3\ \text{joule}$$
$$= 0.252\ \text{kcal}$$
$$= 778\ \text{ft lb}$$
$$= 0.293\ \text{watt hour}$$

$$1\ \text{kcal} = 3.9683\ \text{Btu}$$
$$= 427\ \text{kg m}$$
$$= 4.187 \times 10^3\ \text{joule}$$

$$1\ \text{ft lb} = 0.1383\ \text{kg m}$$
$$= 0.001286\ \text{Btu}$$
$$= 1.356\ \text{joule}$$

$$1\ \text{kg m} = 7.233\ \text{ft lb}$$
$$= 0.00929\ \text{Btu}$$
$$= 9.806\ \text{joule}$$

Conversions *(continued)*

Power

1 watt $= 1$ Nm/s
1 horse power $= 550$ ft lb/s
$= 33,000$ ft lb/m
$= 1.0139$ metric horse power
$= 746$ W
$= 2545$ Btu/h
1 metric horse power $= 736$ W
$= 75$ kg m/s
$= 0.986$ English horse power

Temperatures

$^\circ F = (\frac{9}{5}\,^\circ C)+32$

$^\circ C = \frac{5}{9}(^\circ F - 32)$

1 deg F $= 0.555$ deg C
1 deg C $= 1.8$ deg F

Viscosity

1 poise $= 0.1$ kg/ms
$= 0.1$ N s/m^2
1 stoke $= 1 \times 10^{-4}$ m^2/s

Force

1 N $= 0.2248$ lbf
1 lbf $= 4.448$ N
A mass of 1 kg has a weight of 1 kp
1 kp $= 9.81$ N
Acceleration due to gravity
in London $= 32.2$ ft/s^2
$= 9.81$ m/s^2
at Equator $= 32.1$ ft/s^2
$= 9.78$ m/s^2

Conversion tables

Temperature conversion table. Degrees Fahrenheit to Degrees Centigrade (Figures in italics represent negative values on the Centigrade Scale)

Degrees F	0	1	2	3	4	5	6	7	8	9
	°C	°C	°C	°C	°C	°C	°C	°C	°C	°C
0	*17.8*	*17.2*	*16.7*	*16.1*	*15.6*	*15.0*	*14.4*	*13.9*	*13.3*	*12.8*
10	*12.2*	*11.7*	*11.1*	*10.6*	*10.0*	*9.4*	*8.9*	*8.3*	*7.8*	*7.2*
20	*6.7*	*6.1*	*5.6*	*5.0*	*4.4*	*3.9*	*3.3*	*2.8*	*2.2*	*1.7*
30	*1.1*	*0.6*	–	–	–	–	–	–	–	–
	0	1	2	3	4	5	6	7	8	9
30	–	0	0	0.6	1.1	1.7	2.2	2.8	3.3	3.9
40	4.4	5.0	5.6	6.1	6.7	7.2	7.8	8.3	8.9	9.4
50	10.0	10.6	11.1	11.7	12.2	12.8	13.3	13.9	14.4	15.0
60	15.6	16.1	16.7	17.2	17.8	18.3	18.9	19.4	20.0	20.6
70	21.1	21.7	22.2	22.8	23.3	23.9	24.4	25.0	25.6	26.1
80	26.7	27.2	27.8	28.3	28.9	29.4	30.0	30.6	31.1	31.7
90	32.2	32.8	33.3	33.9	34.4	35.0	35.6	36.1	36.7	37.2
100	37.8	38.3	38.9	39.4	40.0	40.6	41.1	42.7	42.2	42.8
110	43.3	43.9	44.4	45.0	45.6	46.1	46.7	47.2	47.8	48.3
120	48.9	49.4	50.0	50.6	51.1	51.7	52.2	52.8	53.3	53.9
130	54.4	55.0	55.6	56.1	56.7	57.2	57.8	58.3	58.9	59.4
140	60.0	60.6	61.1	61.7	62.2	62.8	63.3	63.9	64.4	65.0
150	65.6	66.1	66.7	67.2	67.8	68.3	68.9	69.4	70.0	70.6
160	71.1	71.7	72.2	72.8	73.3	73.9	74.4	75.0	75.6	76.1
170	76.7	77.2	77.8	78.3	78.9	79.4	80.0	80.6	81.1	81.7
180	82.2	82.8	83.3	83.9	84.4	85.0	85.6	86.1	86.7	87.2
190	87.8	88.3	88.9	89.4	90.0	90.6	91.1	91.7	92.2	92.8
200	93.3	93.9	94.4	95.0	95.6	96.1	96.7	97.2	97.8	98.3
210	98.9	99.4	100.0	100.6	101.1	101.7	102.2	102.8	103.3	103.9
220	104.4	105.0	105.6	106.1	106.7	107.2	107.8	108.3	108.9	109.4
230	110.0	110.6	111.1	111.7	112.2	112.8	113.3	113.9	114.4	115.0
240	115.6	116.1	116.7	117.2	117.8	118.3	118.9	119.4	120.0	120.6
250	121.1	121.7	122.2	122.8	123.3	123.9	124.4	125.0	125.6	126.1

$F = (C \times 1.8) + 32$

Temperature conversion table. Degrees Fahrenheit to Degrees Centigrade *(continued)*

Degrees F	0	1	2	3	4	5	6	7	8	9
	°C	°C	°C	°C	°C	°C	°C	°C	°C	°C
260	126.7	127.2	127.8	128.3	128.9	129.4	130.0	130.6	131.1	131.7
270	132.2	132.8	133.3	133.9	134.4	135.0	135.6	136.1	136.7	137.2
280	137.8	138.3	138.9	139.4	140.0	140.6	141.1	141.7	142.2	142.8
290	143.3	143.9	144.5	145.0	145.6	146.1	146.7	147.2	147.8	148.3
300	148.9	149.4	150.0	150.6	151.1	151.7	152.2	152.8	153.3	153.9
310	154.4	155.0	155.6	156.1	156.7	157.2	157.8	158.3	158.9	159.4
320	160.0	160.6	161.1	161.7	162.2	162.8	163.3	163.9	164.4	165.0
330	165.6	166.1	166.7	167.2	167.8	168.3	168.9	169.4	170.0	170.6
340	171.1	171.7	172.2	172.8	173.2	173.9	174.4	175.0	175.6	176.1
350	176.7	177.2	177.8	178.3	178.9	179.4	180.0	180.6	181.1	181.7
360	182.2	182.8	183.3	183.9	184.4	185.0	185.6	186.1	186.7	187.2
370	187.8	188.3	188.9	189.4	190.0	190.6	191.1	191.7	192.2	192.8
380	193.3	193.9	194.4	195.0	195.6	196.1	196.7	197.2	197.8	198.3
390	198.9	199.4	200.0	200.6	201.1	201.7	202.2	202.8	203.3	203.9
400	204.4	205.0	205.6	206.1	206.7	207.2	207.8	208.3	208.9	209.4
410	210.0	210.6	211.1	211.7	212.2	212.8	213.3	213.9	214.4	215.0
420	215.6	216.1	216.7	217.2	217.8	218.3	218.9	219.4	220.2	220.6
430	221.1	221.7	222.2	222.8	223.3	223.9	224.4	225.0	225.6	226.1
440	226.7	227.2	227.8	228.3	228.9	229.4	230.0	230.6	231.1	231.7
450	232.2	232.8	233.3	233.9	234.4	235.0	235.6	236.1	236.7	237.2
460	237.8	238.3	238.9	239.4	240.0	240.6	241.1	241.7	242.2	242.8
470	243.3	243.9	244.4	245.0	245.6	246.1	246.7	247.2	247.8	248.3
480	248.9	249.4	250.0	250.6	251.1	251.7	252.2	252.8	253.3	253.9
490	254.4	255.0	255.6	256.1	256.7	257.2	257.8	258.3	258.9	259.4
500	260.0	–	–	–	–	–	–	–	–	–

$F = (C \times 1.8) + 32$

Temperature conversion table. Degrees Centigrade to Degrees Fahrenheit

Degrees C	0	1	2	3	4	5	6	7	8	9
	°F	°F	°F	°F	°F	°F	°F	°F	°F	°F
0	32.0	33.8	35.6	37.4	39.2	41.0	42.8	44.6	46.4	48.2
10	50.0	51.8	53.6	55.4	57.2	59.0	60.8	62.6	64.4	66.2
20	68.0	69.8	71.6	73.4	75.2	77.0	78.8	80.6	82.4	84.2
30	86.0	87.8	89.6	91.4	93.2	95.0	96.8	98.6	101.4	102.2
40	104.0	105.8	107.6	109.4	111.2	113.0	114.8	116.6	118.4	120.2
50	122.0	123.8	125.6	127.4	129.2	131.0	132.8	134.6	136.4	138.2
60	140.0	141.8	143.6	145.4	147.2	149.0	150.8	152.6	154.4	156.2
70	158.0	159.8	161.6	163.4	165.2	167.0	168.8	170.6	172.4	174.2
80	176.0	177.8	179.6	181.4	183.2	185.0	186.8	188.6	190.4	192.2
90	194.0	195.8	197.6	199.4	201.2	203.0	204.2	206.6	208.4	210.2
100	212.0	213.8	215.6	217.4	219.2	221.0	222.8	224.6	226.4	228.2
110	230.0	231.8	233.6	235.4	237.2	239.0	240.8	242.6	244.4	246.2
120	248.0	249.8	251.6	253.4	255.2	257.0	258.8	260.6	262.4	264.2
130	266.0	267.8	269.6	271.4	273.2	275.0	276.8	278.6	280.4	282.2
140	284.0	285.8	287.6	289.4	291.2	293.0	294.8	296.6	298.4	300.2
150	302.0	303.8	305.6	307.4	309.2	311.0	312.8	314.6	316.4	318.2
160	320.0	321.8	323.6	325.4	327.2	329.0	330.8	332.6	334.4	336.2
170	338.0	339.8	341.6	343.4	345.2	347.0	348.8	350.6	352.4	354.2
180	356.0	357.8	359.6	361.4	363.2	365.0	366.8	368.6	370.4	372.2
190	374.0	375.8	377.6	379.4	381.2	383.0	384.8	386.6	388.4	390.2
200	392.0	393.8	395.6	397.4	399.2	401.0	402.8	404.6	406.4	408.2
210	410.0	411.8	413.6	415.4	417.2	419.0	420.8	422.6	424.4	426.2
220	428.0	429.8	431.6	433.4	435.2	437.0	438.8	440.6	442.4	444.2
230	446.0	447.8	449.6	451.4	453.2	455.0	456.8	458.6	460.4	462.2
240	464.0	465.8	467.6	469.4	471.2	473.0	474.8	476.6	478.4	480.2
250	482.0	483.8	485.6	487.4	489.2	491.0	492.8	494.6	496.4	498.2
260	500.0	501.8	503.6	505.4	507.2	509.0	510.8	512.6	514.4	516.2
270	518.0	519.8	521.6	523.4	525.2	527.0	528.8	530.6	532.4	534.2
280	536.0	537.8	539.6	541.4	543.2	545.0	546.8	548.6	550.4	552.2
290	554.0	555.8	557.6	559.4	561.2	563.0	563.8	566.6	568.4	570.2
300	572.0	573.8	575.6	577.4	579.2	581.0	582.8	584.6	586.2	588.2

$C = (F - 32) \div 1.8$

2 Standards for materials

Cold water storage and feed and expansion cisterns to BS 417

Imperial sizes

Reference Nos.	Length in	Width in	Depth in	Capacity gal	Thickness Body B.G.	Loose cover B.G.
SC 10	18	12	12	4	16	20
15	24	12	15	8	16	20
20	24	16	15	12	16	20
25	24	17	17	15	16	20
30	24	18	19	19	16	20
40	27	20	20	25	16	20
50	29	22	22	35	14	20
60	30	23	24	42	14	20
70	36	24	23	50	14	20
80	36	26	24	58	14	20
100/2	38	27	27	74	14	20
125	38	30	31	93	12	18
150	43	34	29	108	12	18
200	46	35	35	156	12	18
250	60	36	32	185	12	18
350	60	45	36	270	$\frac{1}{8}$ in	16
500	72	48	40	380	$\frac{1}{8}$ in	16
600	72	48	48	470	$\frac{1}{8}$ in	16
1000	96	60	48	740	$\frac{3}{16}$ in	16

Metric sizes

Reference No.	Length mm	Width mm	Depth mm	Capacity litres	Thickness Body Grade A mm	Grade B mm	Loose cover mm
SCM 45	457	305	305	18	1.6	–	1.0
70	610	305	381	36	1.6	–	1.0
90	610	406	381	54	1.6	–	1.0
110	610	432	432	68	1.6	–	1.0
135	610	457	482	86	1.6	–	1.0
180	686	508	508	114	1.6	–	1.0

Metric sizes *(continued)*

Reference No.	Length mm	Width mm	Depth mm	Capacity litres	Thickness Body Grade A mm	Grade B mm	Loose cover mm
230	736	559	559	159	2.0	1.6	1.0
270	762	584	610	191	2.0	1.6	1.0
320	914	610	584	227	2.0	1.6	1.0
360	914	660	610	264	2.0	1.6	1.0
450/1	1219	610	610	327	2.0	1.6	1.0
450/2	965	686	686	336	2.0	1.6	1.0
570	965	762	787	423	2.5	2.0	1.2
680	1092	864	736	491	2.5	2.0	1.2
910	1168	889	889	709	2.5	2.0	1.2
1130	1524	914	813	841	2.5	2.0	1.2
1600	1524	1143	914	1227	3.2	2.5	1.6
2270	1829	1219	1016	1727	3.2	2.5	1.6
2720	1829	1219	1219	2137	3.2	2.5	1.6
4540	2438	1524	1219	3364	4.8	3.2	1.6

Closed tanks to BS 417

Imperial sizes

Reference No.	Length in	Width in	Depth in	Capacity gal	Thickness in
T25/1	24	17	17	21	$\frac{1}{8}$
25/2	24	24	12	21	$\frac{1}{8}$
30/1	24	18	19	25	$\frac{1}{8}$
30/2	24	24	15	27	$\frac{1}{8}$
40	27	20	20	34	$\frac{1}{8}$

Metric sizes

Reference No.	Length mm	Width mm	Depth mm	Capacity litres	Thickness Grade A mm	Grade B mm
TM114/1	610	432	432	95	3.2	2.5
114/2	610	610	305	95	3.2	2.5
136/1	610	457	482	114	3.2	2.5
136/2	610	610	381	123	3.2	2.5
182	690	508	508	155	3.2	2.5

Copper indirect cylinders to BS 1566:1984

Reference No.	Diameter mm	Height mm	Capacity litres	Heating surface coil m^2
1	350	900	72	0.27
2	400	900	96	0.35
3	400	1050	114	0.42
4	450	675	84	0.31
5	450	750	95	0.35
6	450	825	106	0.40
7	450	900	117	0.44
8	450	1050	140	0.52
9	450	1200	162	0.61
9E	450	1500	206	0.79
10	500	1200	190	0.75
11	500	1500	245	0.87
12	600	1200	280	1.10
13	600	1500	360	1.40
14	600	1800	440	1.70

Copper direct cylinders to BS 699:1984

Reference No.	Diameter mm	Height mm	Capacity litres
1	350	900	74
2	400	900	98
3	400	1050	116
4	450	675	86
5	450	750	98
6	450	825	109
7	450	900	120
8	450	1050	144
9	450	1200	166
9E	450	1500	210
10	500	1200	200
11	500	1500	255
12	600	1200	290
13	600	1500	370
14	600	1800	450

Cold water storage and feed and expansion cisterns of polyolefin or olefin copolymer to BS 4213

Reference no.	Maximum height mm	Capacity litres	Distance of water line from top of cistern mm
PC 4	310	18	110
8	380	36	110
15	430	68	115
20	510	91	115
25	560	114	115
40	610	182	115
50	660	227	115
60	660	273	115
70	660	318	115
100	760	455	115

The standard does not specify width and length.

Sheet and wire gauges

Standard Wire Gauge No.	Birmingham Gauge No.	German Sheet Gauge No. (DIN 1541)	ISO Metric R20 Preferred Series mm	Thickness or Diameter		Weight of Sheet	
				in	mm	lb/ft²	kg/m²
30	–	–	0.315	0.0124	0.315	0.48	2.5
–	–	27	–	0.0126	0.32	0.52	2.5
29	–	–	–	0.0136	0.345	0.52	2.7
–	29	–	–	0.0139	0.354	0.56	2.8
–	–	–	0.355	0.0140	0.355	0.56	2.8
28	–	–	–	0.0148	0.376	0.56	2.9
–	28	–	–	0.0156	0.397	0.63	3.1
–	–	26	–	0.0150	0.38	0.62	3.0
–	–	–	0.400	0.0158	0.400	0.64	3.1
27	–	–	–	0.0164	0.417	0.64	3.2
–	27	–	–	0.0175	0.443	0.71	3.5
–	–	25	–	0.0172	0.44	0.70	3.5
–	–	–	0.450	0.0177	0.450	0.72	3.5
26	–	–	–	0.018	0.457	0.72	3.6
–	26	–	–	0.0196	0.498	0.79	3.9
–	–	24	0.500	0.0197	0.500	0.80	3.9
25	–	–	–	0.020	0.508	0.80	4.0
24	–	–	–	0.022	0.559	0.88	4.4
–	25	–	–	0.022	0.560	0.89	4.4
–	–	23	0.560	0.0221	0.560	0.91	4.4
23	–	–	–	0.024	0.610	1.00	4.8
–	24	–	–	0.025	0.629	1.00	4.9
–	–	22	0.630	0.0248	0.630	1.02	4.9
–	23	–	–	0.028	0.707	1.13	5.5
–	–	–	0.710	0.0280	0.710	1.14	5.6
22	–	–	–	0.028	0.711	1.12	5.6
–	–	21	–	0.0295	0.75	1.21	5.9
–	22	–	–	0.031	0.794	1.27	6.2
–	–	–	0.800	0.0315	0.800	1.28	6.3
21	–	–	–	0.032	0.813	1.28	6.3
–	–	20	–	0.0346	0.88	1.41	6.9
–	21	–	–	0.035	0.887	1.41	7.0
–	–	–	0.900	0.0354	0.900	1.42	7.1
20	–	–	–	0.036	0.914	1.42	7.2
–	20	–	–	0.039	0.996	1.59	7.8
–	–	19	1.000	0.0394	1.000	1.61	7.8
19	–	–	–	0.040	1.016	1.68	8.0
–	19	–	–	0.044	1.12	1.78	8.8
–	–	–	1.12	0.0441	1.12	1.80	8.8
–	–	18	–	0.0443	1.13	1.81	8.9

Sheet and wire gauges *(continued)*

Standard Wire Gauge No.	Birmingham Gauge No.	German Sheet Gauge No. (DIN 1541)	ISO Metric R20 Preferred Series mm	Thickness or Diameter in	mm	Weight of Sheet lb/ft²	kg/m²
18	–	–	–	0.048	1.219	1.96	9.6
–	–	17	1.25	0.0492	1.25	2.00	9.8
–	18	–	–	0.050	1.26	2.00	9.9
–	–	16	–	0.0543	1.38	2.22	10.8
–	–	–	1.40	0.0551	1.40	2.25	11.0
–	17	–	–	0.056	1.41	2.25	11.1
17	–	–	–	0.056	1.422	2.32	11.1
–	–	15	–	0.0591	1.50	2.42	11.7
–	16	–	–	0.063	1.59	2.53	12.4
–	–	–	1.60	0.0630	1.60	2.58	12.5
16	–	–	–	0.064	1.626	2.60	12.7
–	–	14	–	0.0689	1.75	2.82	13.7
–	15	–	–	0.070	1.78	2.83	13.9
–	–	–	1.80	0.0709	1.80	2.90	14.1
15	–	–	–	0.072	1.829	2.94	14.3
–	14	–	–	0.079	1.99	3.18	15.6
–	–	13	2.00	0.0787	2.00	3.18	15.7
14	–	–	–	0.080	2.032	3.32	15.9
–	–	–	–	–	–	–	–
–	13	–	2.24	0.088	2.24	3.57	17.6
–	–	12	–	0.0886	2.25	3.59	17.6
13	–	–	–	0.092	2.337	3.80	18.3
–	–	11	2.50	0.0984	2.50	3.98	19.6
–	12	–	–	0.099	2.52	4.01	19.7
12	–	–	–	0.104	2.642	4.36	20.7
–	–	10	–	0.1083	2.75	4.38	21.6
–	–	–	2.80	0.1102	2.80	4.46	22.0
–	11	–	–	0.111	2.83	4.51	22.2
11	–	–	–	0.116	2.946	4.80	23.1
–	–	9	–	0.1181	3.00	4.56	23.5
–	–	–	3.15	0.1240	3.15	5.02	24.7
–	10	–	–	0.125	3.18	5.06	24.8
–	–	8	–	0.1279	3.25	5.18	25.5
10	–	–	–	0.128	3.251	5.36	25.4
–	–	7	–	0.1378	3.50	5.58	27.4
–	9	–	3.55	0.140	3.55	5.66	27.8
9	–	–	–	0.144	3.658	5.92	28.7
–	–	6	–	0.1476	3.75	5.98	29.4
–	8	–	–	0.157	3.99	6.36	31.3

Sheet and wire gauges (continued)

Standard Wire Gauge No.	Birmingham Gauge No.	German Sheet Gauge No. (DIN 1541)	ISO Metric R20 Preferred Series mm	Thickness or Diameter		Weight of Sheet	
				in	mm	lb/ft²	kg/m²
–	–	5	4.0	0.1575	4.0	6.38	31.4
8	–	–	–	0.160	4.064	6.60	31.9
–	–	4	–	0.1673	4.25	6.77	33.3
7	–	–	–	0.176	4.470	7.12	35.1
–	7	–	–	0.176	4.48	7.14	35.1
–	–	3	4.5	0.1772	4.50	7.17	35.3
6	–	–	–	0.192	4.877	7.80	38.2
–	–	2	5.0	0.1969	5.00	7.97	39.2
–	6	–	–	0.198	5.032	8.02	39.5
5	–	–	–	0.212	5.385	8.80	42.2
–	–	1	–	0.2165	5.50	8.77	43.1
–	–	–	5.6	0.2205	5.6	8.93	43.9
–	5	–	–	0.222	5.66	9.01	44.4
4	–	–	–	0.232	5.893	9.52	46.2
–	–	–	6.30	0.2480	6.30	10.04	49.4
–	4	–	–	0.250	6.35	10.12	49.9
3	–	–	–	0.252	6.401	10.36	50.2
2	–	–	–	0.276	7.010	11.17	55.0
–	–	–	7.10	0.2795	7.10	11.32	55.7
–	3	–	–	0.280	7.13	11.34	55.9
1	–	–	–	0.300	7.620	12.0	59.7
–	2	–	–	0.315	8.00	12.74	62.7
–	–	–	8.00	0.3150	8.00	12.74	62.7
0	–	–	–	0.324	8.229	13.1	63.9
2/0	–	–	–	0.348	8.839	13.9	69.3
–	1	–	–	0.353	8.98	14.30	70.4
–	–	–	9.00	0.3543	9.00	14.3	70.6
3/0	–	–	–	0.372	9.449	14.9	74.1
–	–	–	10.00	0.3937	10.00	15.9	78.4
–	0	–	–	0.396	10.07	16.0	78.9
4/0	–	–	–	0.400	10.160	16.0	79.7
5/0	–	–	–	0.432	10.973	17.3	86.0
–	–	–	11.2	0.4409	11.2	17.8	87.8
–	2/0	–	–	0.445	11.3	18.0	88.6
6/0	–	–	–	0.464	11.785	18.6	92.4
–	–	–	12.5	0.4921	12.5	19.9	98.0
7/0	3/0	–	–	0.500	12.700	20.0	99.5

Weight of steel bar and sheet

Thickness or Diameter mm	Weight in kg of			Thickness or Diameter mm	Weight in kg of		
	Sheet per m²	Square per m	Round per m		Sheet per m²	Square per m	Round per m
5	39.25	0.196	0.154	68	533.80	36.298	28.509
6	47.10	0.283	0.222	70	569.50	36.465	30.210
8	62.80	0.502	0.395	72	585.20	40.694	31.961
10	78.50	0.785	0.617	74	600.90	42.987	33.762
12	94.20	1.130	0.888	76	616.60	45.342	35.611
14	109.90	1.539	1.208	78	632.30	47.759	37.510
16	125.60	2.010	1.578	80	628.00	50.240	39.458
18	141.30	2.543	1.998	85	667.25	56.716	44.545
20	157.00	3.140	2.466	90	706.50	63.585	49.940
22	172.70	3.799	2.984	95	745.75	70.846	55.643
24	188.40	4.522	3.551	100	785.00	78.500	61.654
26	204.10	5.307	4.168	105	824.25	86.546	67.973
28	219.80	6.154	4.834	110	863.5	94.985	74.601
30	235.50	7.065	5.549	115	902.75	103.816	81.537
32	251.20	8.038	6.313	120	942.0	113.040	88.781
34	266.90	9.075	7.127	125	981.2	122.656	96.334
36	282.60	10.174	7.990	130	1020	132.665	104.195
38	298.30	11.335	8.903	135	1060	143.006	112.364
40	314.00	12.560	9.865	140	1099	153.860	120.841
42	329.70	13.847	10.876	145	1138	165.046	129.627
44	345.40	15.198	11.936	150	1178	176.625	138.721
46	361.10	16.611	13.046	155	1217	188.596	148.123
48	376.80	18.086	14.205	160	1256	200.960	157.834
50	392.50	19.625	15.413	165	1295	213.716	167.852
52	408.20	21.226	16.671	170	1355	226.865	178.179
54	423.90	22.891	17.978	175	1394	240.406	188.815
56	439.60	24.618	19.335	180	1413	254.340	199.758
58	455.30	26.407	20.740	185	1452	268.666	211.010
60	471.00	28.260	22.195	190	1492	283.385	222.570
62	486.70	30.175	23.700	195	1511	298.496	234.438
64	502.40	32.154	25.253	200	1570	314.000	246.615
66	518.10	34.195	26.856				

Weight of steel bar and sheet

Thickness or Diameter in	Weight in lb of			Thickness or Diameter in	Weight in lb of		
	Sheet per ft^2	Square per ft	Round per ft		Sheet per ft^2	Square per ft	Round per ft
$\frac{1}{8}$	5.10	0.053	0.042	1	40.80	3.40	2.68
$\frac{3}{16}$	7.65	0.120	0.094	$1\frac{1}{8}$	45.9	4.31	3.38
$\frac{1}{4}$	10.20	0.213	0.167	$1\frac{1}{4}$	51.0	5.32	4.17
$\frac{5}{16}$	12.75	0.332	0.261	$1\frac{3}{8}$	56.1	6.43	5.05
$\frac{3}{8}$	15.30	0.479	0.376	$1\frac{1}{2}$	61.2	7.71	6.01
$\frac{7}{16}$	17.85	0.651	0.511	$1\frac{5}{8}$	66.3	8.99	7.05
$\frac{1}{2}$	20.40	0.851	0.658	$1\frac{3}{4}$	71.4	10.4	8.19
$\frac{9}{16}$	22.95	1.08	0.845	$1\frac{7}{8}$	76.5	12.0	9.39
$\frac{5}{8}$	25.50	1.33	1.04	2	81.6	13.6	10.7
$\frac{11}{16}$	28.05	1.61	1.29	$2\frac{1}{2}$	102.2	21.3	16.8
$\frac{3}{4}$	30.60	1.91	1.50	3	122.4	30.6	24.1
$\frac{13}{16}$	33.15	2.25	1.77	4	163.2	54.4	42.8
$\frac{7}{8}$	35.70	2.61	2.04	5	204.0	85.1	66.9
$\frac{15}{16}$	38.25	2.99	2.35	6	324.8	122.5	96.2

British Standard flanges

Steel flanges to BS 1560 Sect. 3.1: 1989
These are interchangeable with flanges to ANSI B16.5
Class 150

Nominal pipe size in	Outside diameter of flange mm	Diameter of bolt circle mm	No. of bolts	Size of bolts in
$\frac{1}{2}$	89	60.3	4	$\frac{1}{2}$
$\frac{3}{4}$	98	69.8	4	$\frac{1}{2}$
1	108	79.4	4	$\frac{1}{2}$
$1\frac{1}{2}$	127	98.4	4	$\frac{1}{2}$
2	152	120.6	4	$\frac{5}{8}$
$2\frac{1}{2}$	178	139.7	4	$\frac{5}{8}$
3	190	152.4	4	$\frac{5}{8}$
4	229	190.5	8	$\frac{5}{8}$
6	279	241.3	8	$\frac{3}{4}$
8	343	298.4	8	$\frac{3}{4}$
10	406	362.0	12	$\frac{7}{8}$
12	483	431.8	12	$\frac{7}{8}$
14	533	476.2	12	1
16	597	539.8	16	1
18	635	577.8	16	$1\frac{1}{8}$
20	698	635.0	20	$1\frac{1}{8}$
24	813	749.3	20	$1\frac{1}{4}$

British Standard flanges

Steel flanges to BS 1560 Sect. 3.1: 1989

These are interchangeable with flanges to ANSI B16.5
Class 300

Nominal pipe size in	Outside diameter of flange mm	Diameter of bolt circle mm	No. of bolts	Size of bolts in
$\frac{1}{2}$	95	66.7	4	$\frac{1}{2}$
$\frac{3}{4}$	117	82.6	4	$\frac{5}{8}$
1	124	88.9	4	$\frac{5}{8}$
$1\frac{1}{2}$	156	114.3	4	$\frac{3}{4}$
2	165	127.0	8	$\frac{5}{8}$
$2\frac{1}{2}$	190	149.2	8	$\frac{3}{4}$
3	210	168.3	8	$\frac{3}{4}$
4	254	200.0	8	$\frac{3}{4}$
6	318	269.9	12	$\frac{3}{4}$
8	381	330.2	12	$\frac{7}{8}$
10	444	387.4	16	1
12	521	450.8	16	$1\frac{1}{8}$
14	584	514.4	20	$1\frac{1}{8}$
16	648	571.5	20	$1\frac{1}{4}$
18	711	628.6	24	$1\frac{1}{4}$
20	775	685.8	24	$1\frac{1}{4}$
24	914	812.8	24	$1\frac{1}{2}$

Metric pipe flanges to BS 4504

Nominal pressure - 2.5 bar
Thickness of flange depends on type and material

Nominal pipe size	Outside diameter of pipe mm	Diameter of flange mm	Diameter of bolt circle mm	No. of bolts	Size of bolts
10	17.2	75	50	4	M10
15	21.3	80	55	4	M10
20	26.9	90	65	4	M10
25	33.7	100	75	4	M10
32	42.4	120	90	4	M12
40	48.3	130	100	4	M12
50	60.3	140	110	4	M12
65	76.1	160	130	4	M12
80	88.9	190	150	4	M16
100	114.3	210	170	4	M16
125	139.7	240	200	8	M16
150	168.3	265	225	8	M16
200	219.1	320	280	8	M16
250	273	375	335	12	M16
300	323.9	440	395	12	M20
350	355.6	490	445	12	M20
400	406.4	540	495	16	M20
500	508	645	600	20	M20
600	609.6	755	705	20	M24

Nominal pressure - 6 bar
Dimensions as for 2.5 bar for sizes up to 600 NB

Metric pipe flanges to BS 4504

Nominal pressure - 10 bar
Thickness of flange depends on type and material

Nominal pipe size	Outside diameter of pipe mm	Diameter of flange mm	Diameter of bolt circle mm	No. of bolts	Size of bolts
10	17.2	90	60	4	M12
15	21.3	95	65	4	M12
20	26.9	105	75	4	M12
25	33.7	115	85	4	M12
32	42.4	140	100	4	M16
40	48.3	150	110	4	M16
50	60.3	165	125	4	M16
65	76.1	185	145	4	M16
80	88.9	200	160	8	M16
100	114.3	220	180	8	M16
125	139.7	250	210	8	M16
150	168.3	285	240	8	M20
200	219.1	340	295	8	M20
250	273	395	350	12	M20
300	323.9	445	400	12	M20
350	355.6	505	460	16	M20
400	406.4	565	515	16	M24
500	508	670	620	20	M24
600	609.6	780	725	20	M27

Metric pipe flanges to BS 4504

Nominal pressure – 16 bar
Thickness of flange depends on type and material

Nominal pipe size	Outside diameter of pipe mm	Diameter of flange mm	Diameter of bolt circle mm	No. of bolts	Size of bolts
10	17.2	90	60	4	M12
15	21.3	95	65	4	M12
20	26.9	105	75	4	M12
25	33.7	115	85	4	M12
32	42.4	140	100	4	M16
40	48.3	150	110	4	M16
50	60.3	165	125	4	M16
65	76.1	185	145	4	M16
80	88.9	200	160	8	M16
100	114.3	220	180	8	M16
125	139.7	250	210	8	M16
150	168.3	285	240	8	M20
200	219.1	340	295	12	M20
250	273	405	355	12	M24
300	323.9	460	410	12	M24
350	355.6	520	470	16	M24
400	406.4	580	525	16	M27
500	508	715	650	20	M30
600	609.6	840	770	20	M33

Metric pipe flanges to BS 4504

Nominal pressure – 25 bar
Thickness of flange depends on type and material

Nominal pipe size	Outside diameter of pipe mm	Diameter of flange mm	Diameter of bolt circle mm	No. of bolts	Size of bolts
10	17.2	90	60	4	M12
15	21.3	95	65	4	M12
20	26.9	105	75	4	M12
25	33.7	115	85	4	M12
32	42.4	140	100	4	M16
40	48.3	150	110	4	M16
50	60.3	165	125	4	M16
65	76.1	185	145	8	M16
80	88.9	200	160	8	M16
100	114.3	235	190	8	M20
125	139.7	270	220	8	M24
150	168.3	300	250	8	M24
200	219.1	360	310	12	M24
250	273	425	370	12	M27
300	323.9	485	430	16	M27
350	355.6	555	490	16	M30
400	406.4	620	550	16	M33
500	508	730	660	20	M33

Dimensions of tubes

General dimensions of steel tubes to BS 1387: 1985
(Subject to standard tolerances and usual working allowances)

Nominal bore		Outside diameter		Thickness			Mass of black tube					
in	mm	Light	Heavy & Medium	Light	Medium	Heavy	Light		Medium		Heavy	
							Plain	Screwed	Plain	Screwed	Plain	Screwed
	mm	mm	mm	mm	mm	mm	kg/m	kg/m	kg/m	kg/m	kg/m	kg/m
1/4	8	13.6	13.9	1.8	2.3	2.9	0.515	0.519	0.641	0.645	0.765	0.769
3/8	10	17.1	17.4	1.8	2.3	2.9	0.670	0.676	0.839	0.845	1.02	1.03
1/2	15	21.4	21.7	2.0	2.6	3.2	0.947	0.956	1.21	1.22	1.44	1.45
3/4	20	26.9	27.2	2.3	2.6	3.2	1.38	1.39	1.56	1.57	1.87	1.88
1	25	33.8	34.2	2.6	3.2	4.0	1.98	2.00	2.41	2.43	2.94	2.96
1 1/4	32	42.5	42.9	2.6	3.2	4.0	2.54	2.57	3.10	3.13	3.80	3.83
1 1/2	40	48.4	48.8	2.9	3.2	4.0	3.23	3.27	3.57	3.61	4.38	4.42
2	50	60.2	60.8	2.9	3.6	4.5	4.08	4.15	5.03	5.10	6.19	6.26
2 1/2	65	76.0	76.6	3.2	3.6	4.5	5.71	5.83	6.43	6.55	7.93	8.05
3	80	88.7	89.5	3.2	4.0	5.0	6.72	6.89	8.37	8.54	10.3	10.5
4	100	113.9	114.9	3.6	4.5	5.4	9.75	10.0	12.2	12.5	14.5	14.8
5	125	–	140.6	–	5.0	5.4	–	–	16.6	17.1	17.9	18.4
6	150	–	166.1	–	5.0	5.4	–	–	19.7	20.3	21.3	21.9

Suggested maximum working pressures

The pressures given below can be taken as conservative estimates for tubes screwed taper with sockets tapped parallel under normal (non-shock) conditions

	Grade	Nom. bore	$\frac{1}{8}$ to 1 in	$1\frac{1}{4}$ & $1\frac{1}{2}$ in	2 & $2\frac{1}{2}$ in	3 in	4 in	5 in	6 in
Water	light	lb/in^2	150	125	100	100	80	–	–
		kN/m^2	1000	850	700	700	550	–	–
	medium	lb/in^2	300	250	200	200	150	150	125
		kN/m^2	2000	1750	1400	1400	1000	1000	850
	heavy	lb/in^2	350	300	250	250	200	200	150
		kN/m^2	2400	2000	1750	1750	1400	1400	1000
Steam or air	medium	lb/in^2	150	125	100	100	80	80	60
		kN/m^2	1000	850	700	700	550	550	400
	heavy	lb/in^2	175	150	125	125	100	100	80
		kN/m^2	1200	1000	850	850	700	700	550

The following allowed for plain end tubes end-to-end welded for steam or compressed air.

medium	lb/in^2	250	200	200	150	150	150	125
	kN/m^2	1750	1400	1400	1000	1000	1000	850
heavy	lb/in^2	300	300	300	200	200	200	75
	kN/m^2	2000	2000	2000	1400	1400	1400	1200

Copper tube to BS 2871: 1972

Nominal bore mm	Outside diameter mm	Table X Half hard light gauge tube		Table Y Half hard and annealed tube		Table Z Hard drawn thin wall tube	
		Thickness mm	Maximum working pressure N/mm^2	Thickness mm	Maximum working pressure N/mm^2	Thickness mm	Maximum working pressure N/mm^2
6	6	0.6	13.3	0.8	14.4	0.5	11.3
8	8	0.6	9.7	0.8	10.5	0.5	9.8
10	10	0.6	7.7	0.8	8.2	0.5	7.8
12	12	0.6	6.3	0.8	6.7	0.5	6.4
15	15	0.7	5.8	1.0	6.7	0.5	5.0
18	18	0.8	5.6	1.0	5.5	0.6	5.0
22	22	0.9	5.1	1.2	5.7	0.6	4.1
28	28	0.9	4.0	1.2	4.2	0.6	3.2
35	35	1.2	4.2	1.5	4.1	0.7	3.0
42	42	1.2	3.5	1.5	3.4	0.8	2.8
54	54	1.2	2.7	2.0	3.6	0.9	2.5
67	67	1.2	2.0	2.0	2.8	1.0	2.0
76.1	76.2	1.5	2.4	2.0	2.5	1.2	1.9
108	108.1	1.5	1.7	2.5	2.2	1.2	1.7
133	133.4	1.5	1.4	–	–	1.5	1.6
159	159.4	2.0	1.5	–	–	1.5	1.5

Malleable iron pipe fittings

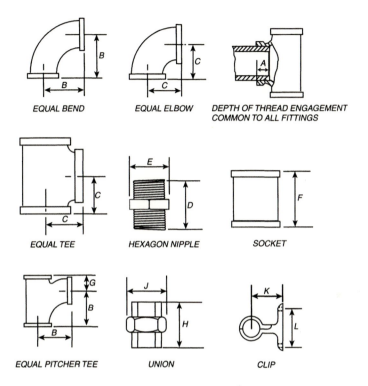

EQUAL BEND

EQUAL ELBOW

DEPTH OF THREAD ENGAGEMENT
COMMON TO ALL FITTINGS

EQUAL TEE

HEXAGON NIPPLE

SOCKET

EQUAL PITCHER TEE

UNION

CLIP

Dimensions of malleable iron pipe fittings
Dimensions in mm

Nominal bore		*15*	*20*	*25*	*32*	*40*	*50*	*65*	*80*	*100*
Depth of thread engagement	A	13	14	16	19	19	24	25	29	35
Bend	B	45	50	63	76	85	102	114	127	165
Elbow	C	28	33	38	45	50	58	69	78	96
Equal Tee	C	28	33	38	45	50	58	69	78	96
Hexagon nipple length	D	44	49	56	64	64	71	80	89	102
across flats	E	23	28	35	44	50	61	77	90	115
Socket length	F	34	39	42	49	54	64	73	81	94
Equal pitcher Tee	B	45	50	63	76	85	102	114	127	165
	G	24	28	33	40	43	53	61	70	87
Union length	H	46	52	57	64	68	75	84	92	106
across flats	J	42	48	57	68	76	92	109	125	155
Pipe clip	K	43	43	51	56	70	76	89	97	118
	L	40	48	54	60	73	86	95	108	143

3 Combustion

Atomic weights of elements occurring in combustion calculations

Element	Symbol	Atomic No.	Atomic weight
Carbon	C	6	12.011
Hydrogen	H	1	1.008
Nitrogen	N	7	14.007
Oxygen	O	8	15.9994
Phosphorus	P	15	30.9738
Sulphur	S	16	32.06

Heat of combustion of important chemicals

Substance	Products of combustion	Chemical equation	Heat of combustion kJ/kg	Heat of combustion Btu/lb
Carbon	Carbon dioxide	$C + O_2 = CO_2$	33,950	14,590
Carbon	Carbon monoxide	$2C + O_2 = 2CO$	9,210	3,960
Carbon monoxide	Carbon dioxide	$2CO + O_2 = 2CO_2$	10,150	4,367
Hydrogen	Water	$2H_2 + O_2 = 2H_2O$	144,200	62,000
Sulphur	Sulphur dioxide	$S + O_2 = SO_2$	9,080	3,900
Methane	Carbon dioxide and water	$CH_4 + 2O_2 = CO_2 + 2H_2O$	55,860	24,017

Ignition temperatures

Wood	300°C	570°F	Petroleum	400°C	750°F
Peat	227°C	440°F	Benzene	415°C	780°F
Bituminous coal	300°C	570°F	Coal-tar oil	580°C	1080°F
Semi anthracite coal	400°C	750°F	Producer gas	750°C	1380°F
Coke	700°C	1290°F	Light hydrocarbons	650°C	1200°F
Hydrogen	500°C	930°F	Heavy hydrocarbons	750°C	1380°F
Carbon monoxide	300°C	570°F	Light gas	600°C	1110°F
Carbon	700°C	1290°F	Naphtha	550°C	1020°F

Composition and calorific value of fuels

Fuel	Composition by weight						Higher calorific value	
	C	H	O+N	S	H_2O	Ash	kJ/kg	Btu/lb
Anthracite	83–87	3.5–4.0	3.0–4.7	0.9	1–3	4–6	32 500–34 000	14 000–14 500
Semi-anthracite	63–76	3.5–4.8	8–10	0.5–1.8	5–15	4–14	26 700–32 500	11 500–14 000
Bituminous coal	46–56	3.5–5.0	9–16	0.2–3.0	18–32	2–10	17 000–23 250	73 00–10 000
Lignite	37	7	13.5	0.5	37	5	16 300	7000
Peat	38–49	3.0–4.5	19–28	0.2–1.0	16–29	1–9	13 800–20 500	5500–8800
Coke	80–90	0.5–1.5	1.5–5.0	0.5–1.5	1–5	5–12	28 000–31 000	12 000–13 500
Charcoal	84	1	–	–	12	3	29 600	12 800
Wood (dry)	35–45	3.0–5.0	34–42	–	7–22	0.3–3.0	14 400–17 400	6200–7500
							kJ/m^3	Btu/ft^3
Town gas	26	56	18	–	–	–	18 600	500
Natural gas	75	25	–	–	–	–	37 200	1000
Propane C_3H_8	82	18	–	–	–	–	93 900	2520
Butane C_4H_{10}	83	17	–	–	–	–	130 000	3490
							kJ/l	Btu/gal
Kerosine	86.2	13.0	–				35 000	154 000
Gas oil	85.0	10.8	–	0.8			38 000	164 000
Heavy fuel oil			–	3.8			41 200	177 000

THE RINGELMANN SCALE FOR GRADING DENSITY OF SMOKE

SMOKE NUMBER	0	1	2	3	4	5
LINES mm / SPACES mm	ALL WHITE	1 9	2.3 7.7	3.7 6.3	5.5 4.5	ALL BLACK

Observer should stand 30–300 m from stack and hold scale at arm's length. He should then determine the shade in the chart most nearly corresponding to the shade of the smoke. Care should be taken to avoid either bright sunlight or dark buildings in the background.

Excess of air for good conditions

For anthracite and coke	40%
For semi-anthracite, hand firing	70–100%
For semi-anthracite, with stoker	40–70%
For semi-anthracite, with travelling grate	30–60%
For oil	10–20%
For gas	10%

Theoretical values of combustion air and flue gases

Fuel	Theoretical air for combustion Volume at S.T.P.		Theoretical flue gas produced Volume at S.T.P.	
	m^3/kg	ft^3/lb	m^3/kg	ft^3/lb
Anthracite	9.4	150	9.5	152
Semi-anthracite	8.4	135	8.6	137
Bituminous coal	6.9	110	7.0	112
Lignite	5.7	92	5.8	93
Peat	5.7	92	5.9	94
Coke	8.4	134	8.4	135
Charcoal	8.4	134	8.4	135
Wood (dry)	4.4	70	5.0	80
	$m^3 air/m^3 fuel$	$ft^3 air/ft^3 fuel$	$m^3 gas/m^3 fuel$	$ft^3 gas/ft^3 fuel$
Town gas	4	4	3.8	3.8
Natural gas	9.5	9.5	8.5	8.5
Propane C_3H_8	24.0	24.0	22	22
Butane C_4H_{10}	31	31	27	27
	$m^3 air/litre fuel$	$ft^3 air/gal fuel$	$m^3 gas/litre fuel$	$ft^3 gas/gal fuel$
Gas oil	9.8	1570	10.4	1670
Heavy fuel oil	10.8	1730	11.6	1860

Heat losses in a boiler

1   Sensible heat carried away by dry flue gases

$$L_1 = WC_p(t_1 - t_A) \text{ kJ per kg of fuel}$$
$$= WC_p(t_1 - t_A)\frac{100}{S} \text{ per cent}$$

2   Heat lost by free moisture in fuel

$$L_2 = w(H - h) \text{ kJ per kg of fuel}$$
$$= w(H - h)\frac{100}{S} \text{ per cent}$$

3   Heat lost by incomplete combustion

$$L_3 = 24\,000 \frac{CO}{CO_2 + CO} C \text{ kJ per kg of fuel}$$
$$= 24\,000 \frac{CO}{CO_2 + CO} C \times \frac{100}{S} \text{ per cent}$$

4   Heat lost due to Carbon in Ash

$$L_4 = W_c \times 33\,950 \text{ kJ per kg of fuel}$$
$$= W_c \times 33\,950 \times \frac{100}{S} \text{ per cent}$$

5   Heat lost by Radiation and Unaccounted Losses obtained by difference

$$L_s = S - (M + L_1 + L_2 + L_3 + L_4)$$

where

W = weight of combustion products, kg per kg fuel
W_c = weight of carbon in ash, kg per kg fuel
w = weight of water in fuel, kg per kg fuel
C_p = specific heat capacity of flue gas, kg per kJ per deg C
   = 1.0
t_1 = temperature of flue gas °C
t_A = ambient temperature in boiler room, °C
S = higher calorific value of fuel, kJ per kg
H = total heat of superheated steam at temperature t_1 and atmospheric pressure, kJ per kg
h = sensible heat of water at temperature t_A, kJ per kg
C = weight of carbon in fuel, kg per kg
CO = percentage by volume of carbon monoxide in flue gas
CO_2 = percentage by volume of carbon dioxide in flue gas
M = utilised heat in boiler output.

The largest loss is normally the sensible heat in the flue gases. In good practice it is about 20%.

Sensible heat carried away by flue gases

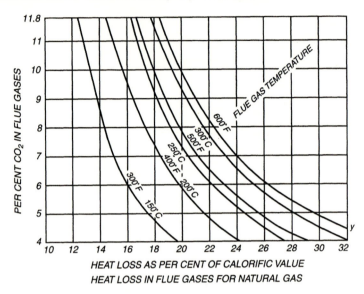

HEAT LOSS AS PER CENT OF CALORIFIC VALUE

HEAT LOSS IN FLUE GASES FOR NATURAL GAS

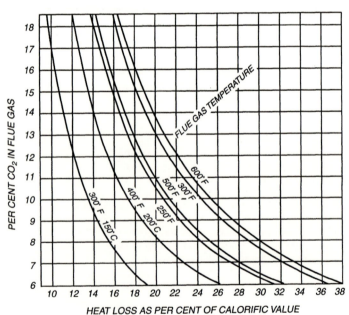

HEAT LOSS AS PER CENT OF CALORIFIC VALUE

HEAT LOSS IN FLUE GASES FOR TOWN GAS

Sensible heat carried away by flue gases

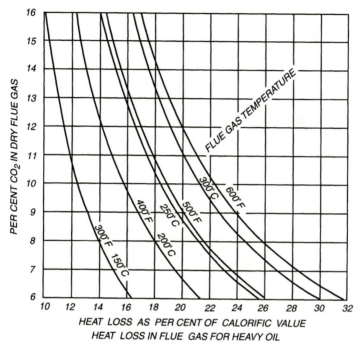

HEAT LOSS AS PER CENT OF CALORIFIC VALUE
HEAT LOSS IN FLUE GAS FOR HEAVY OIL

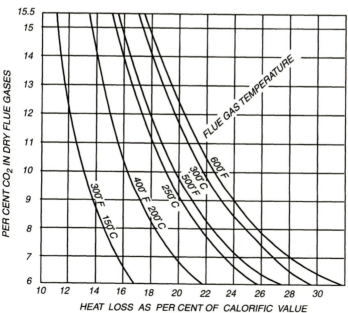

HEAT LOSS AS PER CENT OF CALORIFIC VALUE
HEAT LOSS IN FLUE GAS FOR GAS OIL

Chimney sizes

Theoretical chimney draught

$$h = 354\,H\left(\frac{1}{T_1} - \frac{1}{T_2}\right)$$

where

h = draught in mm water
H = chimney height in m
T_1 = absolute temperature outside chimney K
T_2 = absolute temperature inside chimney K

$$h = 7.64\,H\left(\frac{1}{T_1} - \frac{1}{T_2}\right)$$

where

h = draught in inches of water
H = chimney height in ft
T_1 = absolute temperature outside chimney $°R$
T_2 = absolute temperature inside chimney $°R$.

Chimney area

The chimney should be designed to give a maximum velocity of 2 m/s (7 ft/s) for small furnaces, and 10–15 m/s for large furnaces.

$$A = \frac{Q}{V}$$

where

A = cross-sectional area of chimney, m^2
Q = volume of flue gases at chimney temperature, m^3/s
V = velocity, m/s

An empirical rule is to provide 400 mm^2 chimney area per 1 kW boiler rating (0.2 in^2 per 1000 Btu/hour boiler rating).

Recommended sizes of explosion doors or draught stabilisers for oil firing installations

Cross-sectional area of chimney in^2	Release opening of stabiliser, approx., in	Cross-sectional area of chimney m^2	Release opening of stabiliser, approx., mm
40–80	6×9	0.025–0.050	150×230
80–200	8×13	0.050–0.125	200×330
200–300	13×18	0.125–0.200	330×450
300–600	16×24	0.200–0.400	400×600
600–1500	24×32	0.400–1.000	600×800

Combustion air

A boiler house must have openings to fresh air to allow combustion air to enter. An empirical rule is to allow 1600 mm^2 free area per 1 kW boiler rating (1.5 in^2 per 2000 Btu/hour boiler rating).

Flue dilution

Flue gases from gas burning appliances can be diluted with fresh air to enable the products of combustion to be discharged at low level or near windows.

Typical arrangements

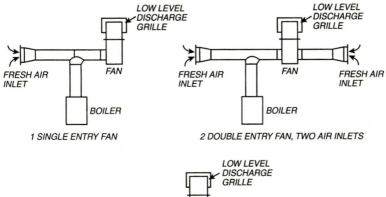

1 SINGLE ENTRY FAN   2 DOUBLE ENTRY FAN, TWO AIR INLETS

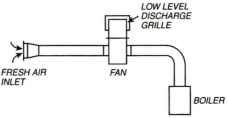

3 DOUBLE ENTRY FAN, SINGLE AIR INLET

To reduce CO_2 concentration to 1%, fan must handle 100 m^3 mixed volume per 1 m^3 natural gas fuel burnt.

In determining fan pressure allowance must be made for pressure due to local wind conditions.

Discharge grille must have free area not less than that of flue.

Fresh air intake must have free area not less than that of flue.

Fresh air intake should be on same face of building as discharge grille in order to balance out wind effect.

Density and specific volume of stored fuels

Fuel	Density		Specific volume	
	kg/m^3	lb/ft^3	m^3 per 1000 kg	ft^3 per ton
Wood	360–385	22.5–2.4	2.5–2.8	90–100
Charcoal, hard wood	149	9.3	6.7	240
Charcoal, soft wood	216	13.5	4.6	165
Anthracite	720–850	45–53	1.2–1.4	42–50
Bituminous coal	690–800	43–50	1.2–1.5	45–52
Peat	310–400	19.5–25	2.5–3.2	90–115
Coke	375–500	23.5–31	2.0–2.7	72–95
Kerosine	790	49	1.3	47
Gas oil	835	52	1.2	43
Fuel oil	930	58	1.1	39

Classification of Oil Fuels
Based on BS 2869

Common Name	Kerosine	Gas oil	Fuel oil or heavy fuel oil			
Class to BS 2869	C1	C2	D	E	F	G
Kinematic viscosity						
cSt at 40°C	–	1.0–2.0	1.5–5.5			
cSt at 100°C	–			8.2 max.	20 max.	40 max.
Flash point, closed						
Abel, min. °C	43	38	56			
Pensky-Martin, min. °C	–			66	66	66
Sulphur content per cent by mass	0.04	0.2	0.5	3.5	3.5	3.5
Minimum temperature for storage °C	ambient	ambient	ambient	10	25	40
for outflow from storage and handling °C	ambient	ambient	ambient	10	30	50
Application	Distillate fuel for free standing flue less domestic appliances	Similar for vapourising and atomising burners on domestic appliances with flues	Distillate fuel for atomising burners for domestic and industrial use	Residual or blended fuels for atomising burners normally requiring preheating before combustion in burner		

Classification of Coal

Based on volatile matter and coking power of clean material

Class	Volatile matter percent (dry mineral matter free basis)	General description	
101	<6.1	Anthracites	
102	6.1–9.0		
201	9.1–13.5	Dry steam coals	Low volatile steam coals
202	13.6–15.0		
203	15.1–17.0	Coking steam coals	
204	17.1–19.5		
206	9.1–19.5	Heat altered low volatile steam coals	
301	19.6–32.0	Prime coking coals	Medium volatile coals
305	19.6–32.0	Mainly heat altered	
306	19.6–32.0	coals	
401	32.1–36.0	Very strongly	
402	>36.0	coking coals	
501	32.1–36.0	Strongly coking	
502	>36.0	coals	
601	32.1–36.0	Medium coking	High volatile coals
602	>36.0	coals	
701	32.1–36.0	Weakly coking	
702	>36.0	coals	
801	32.1–36.0	Very weakly coking	
802	>36.0	coals	
901	32.1–36.0	Non-coking coals	
902	>36.0		

Flow of oil in pipes

Head loss of various viscosities for laminar flow

Viscosity at temp. in pipe cS	4.0	25	45	250	500
i_1	$0.54 \times 10^{-4} \dfrac{f_1}{d_1^4}$	$3.4 \times 10^{-4} \dfrac{f_1}{d_1^4}$	$6.1 \times 10^{-4} \dfrac{f_1}{d_1^4}$	$34 \times 10^{-4} \dfrac{f_1}{d_1^4}$	$67 \times 10^{-4} \dfrac{f_1}{d_1^4}$
i_2	$1.7 \times 10^4 \dfrac{f_2}{d_2^4}$	$11 \times 10^4 \dfrac{f_2}{d_2^4}$	$20 \times 10^4 \dfrac{f_2}{d_2^4}$	$110 \times 10^4 \dfrac{f_2}{d_2^4}$	$220 \times 10^4 \dfrac{f_2}{d_2^4}$

$i_1 = i_2 =$ head loss in feet of oil per foot of pipe or metres of oil per metre
of pipe. (Length of pipe to include allowances for bends, valves
and fittings)
$f_1 =$ flow of oil in gal/hr
$f_2 =$ flow of oil in litre/s
$d_1 =$ internal diameter of pipe in inches
$d_2 =$ internal diameter of pipe in mm.

The above formulae are for laminar flow. Flow is laminar if Reynolds
Number (Re) is less than 1500. Reynolds number can be checked from the following formulae. As Re is a dimensionless ratio it is the same in all consistent systems of units. The coefficients in the following formulae take into
account the dimensions of f_1, d_1, f_2, d_2 respectively.
The viscosity to be taken is that at the temperature of the oil in the pipe.

Viscosity at temp. in pipe cS	4.0	25	45	250	500
Re	$16 \dfrac{f_1}{d_1}$	$2.5 \dfrac{f_1}{d_1}$	$1.0 \dfrac{f_1}{d_1}$	$0.25 \dfrac{f_1}{d_1}$	$0.12 \dfrac{f_1}{d_1}$
	$32 \times 10^4 \dfrac{f_2}{d_2}$	$4.5 \times 10^4 \dfrac{f_2}{d_2}$	$2.8 \times 10^4 \dfrac{f_2}{d_2}$	$0.45 \times 10^4 \dfrac{f_2}{d_2}$	$0.25 \times 10^4 \dfrac{f_2}{d_2}$

Heat loss from oil tanks

| | Oil temperature | | Heat loss | | | |
| | | | Unlagged | | Lagged | |
Position	°F	°C	$\frac{Btu}{ft^2\,hr\,°F}$	$\frac{W}{m^2\,K}$	$\frac{Btu}{ft^2\,hr\,°F}$	$\frac{W}{m^2\,K}$
Sheltered	up to 50	up to 10	1.2	6.8	0.3	1.7
	50–80	10–27	1.3	7.4	0.325	1.8
	80–100	27–38	1.4	8.0	0.35	2.0
Exposed	up to 50	up to 10	1.4	8.0	0.35	2.0
	50–80	10–27	1.5	8.5	0.375	2.1
	80–100	27–38	1.6	9.0	0.4	2.25
In pit			Nil		Nil	

Heat transfer coefficients for coils are:

Steam to oil: 11.3 W/m^2 °C   20 Btu/ft^2 hr °F
Hot water to oil: 5.7 W/m^2 °C   10 Btu/ft^2 hr °F

Heat loss from oil pipes

| Nominal bore | Oil temperature | | Heat loss | |
| | | | Btu | W |
mm	°F	°C	$\frac{Btu}{hr\,ft\,°F}$	$\frac{W}{mK}$
15	up to 50	up to 10	0.4	0.7
20			0.4	0.7
25			0.8	1.4
40			1.2	2.1
50			1.6	2.8
15	50–80	10–27	0.5	0.9
20			0.6	1.1
25			0.7	1.2
40			1.0	1.7
50			1.2	2.1
15	80–100	27–38	0.5	0.9
20			0.6	1.1
25			0.8	1.4
40			1.1	1.9
50			1.3	2.2

Diagrammatic arrangement of oil storage tank

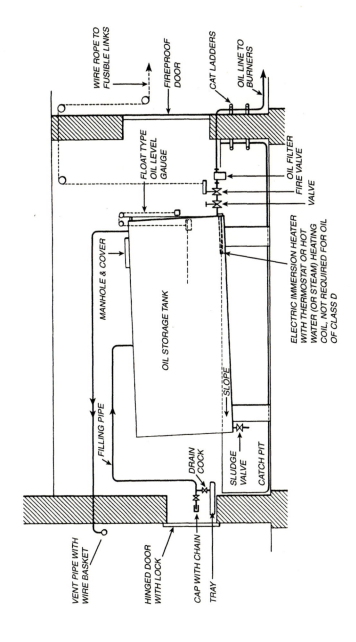

WIRE ROPE TO FUSIBLE LINKS

FIREPROOF DOOR

CAT LADDERS

OIL LINE TO BURNERS

FLOAT TYPE OIL LEVEL GAUGE

OIL FILTER

FIRE VALVE

VALVE

MANHOLE & COVER

ELECTRIC IMMERSION HEATER WITH THERMOSTAT OR HOT WATER (OR STEAM) HEATING COIL, NOT REQUIRED FOR OIL OF CLASS D

OIL STORAGE TANK

SLOPE

FILLING PIPE

DRAIN COCK

SLUDGE VALVE

CATCH PIT

VENT PIPE WITH WIRE BASKET

HINGED DOOR WITH LOCK

CAP WITH CHAIN

TRAY

Diagrammatic arrangement of oil storage tank and day oil tank

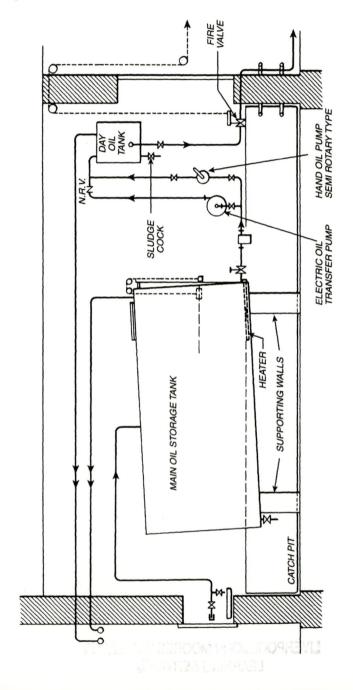

4 Heat and thermal properties of materials

Expansion by heat

Linear expansion is the increase in length

$$L_2 = L_1(1+et)$$

Surface expansion is the increase in area

$$A_2 = A_1(1+2et)$$

Volumetric expansion is the increase in volume

$$V_2 = V_1(1+3et)$$

where

t = temperature difference (K)
L_1 = original length (m)
A_1 = original area (m^2)
V_1 = original volume (m^3)
L_2 = final length (m)
A_2 = final area (m^2)
V_2 = final volume (m^3)
e = coefficient of linear expansion (m/mK)

Sensible heat for heating or cooling

$$H = cM(t_2 - t_1)$$

where

H = Heat (J)
M = mass (kg)
c = specific heat capacity (J/kg K)
t_1 = initial temperature (°C)
t_2 = final temperature (°C)

Expansion of gases

General gas law

$$PV = mRT$$
$$R = (C_p - C_v)$$

where

P = pressure (absolute), N/m^2 (lbf/ft^2)
V = volume, m^3 (ft^3)
m = mass, kg (lbm)
R = gas constant, J/kg K (ft lbf/lbm K)
T = absolute temperature, °K (°R)
C_p = specific heat capacity at constant pressure, J/kg K (Btu/lb°F)
C_v = specific heat capacity at constant volume, J/kg K (Btu/lb°F)

For air

R = 287 J/kg K
  = 96 ft lbf/lbm K
  = 53.3 ft lbf/lbm °F

$$G = MR = \text{universal gas constant which is the same for all gases}$$

where

G = universal gas constant J/kg K
M = molecular weight of gas (dimensionless)
R = gas constant for the gas J/kg K

$$G = (Cp_m - Cv_m)$$

where

Cp_m = specific heat capacity at constant pressure in J/kg mol K
Cv_m = specific heat capacity at constant volume in J/kg mol K

$$PV_m = nGT$$

where

V_m = volume of n moles
n = number of moles

In various units

G = 1.985 Btu/lb °F
  = 1544 ft lbf/lbm °F
  = 2780 ft lbf/lbm K
  = 1.985 kcal/kg K
  = 8.314 kJ/kg K

At N.T.P. 1 kg mol occupies 22.4 m^3
          1 lb mol occupies 359 ft^3

At S.T.P. 1 kg mol occupies 23.7 m^3
          1 lb mol occupies 379 ft^3

Methods of heating and expanding gases (not vapours)

Type of expansion	Remarks	Work done W	Change of internal energy E	Heat absorbed H	Final temperature
Constant pressure	Isobar	$P(V_2 - V_1)$	$MC_v(T_2 - T_1)$	$MC_p(T_2 - T_1)$	$T_1\left(\dfrac{V_2}{V_1}\right)$
Constant temperature	Isotherm	$P_1V_1\log_e\dfrac{V_2}{V_1}$	0	$P_1V_1\log_e\dfrac{V_2}{V_1}$	T_1
Constant heat	Adiabatic $PV^\gamma = $ const.	$\dfrac{P_1V_1 - P_2V_2}{\gamma - 1}$	$MC_v(T_2 - T_1)$	0	$T_1\left(\dfrac{V_1}{V_2}\right)^{\gamma-1}$
Int. energy & temperature change	Polytrope $PV^n = $ const.	$\dfrac{P_1V_1 - P_2V_2}{n - 1}$	$MC_v(T_2 - T_1)$	$W + E$	$T_1\left(\dfrac{V_1}{V_2}\right)^{n-1}$

where

W = external work done by gas (kJ)
E = increase of internal energy by gas (kJ)
H = total heat absorbed (kJ)
P_1, P_2 = initial, final, pressure (N/m²)
V_1, V_2 = initial, final, volume (m³)

T_1, T_2 = initial, final, temperature (°C)
M = mass (kg)
$\gamma = C_p/C_v$ (dimensionless)
n = index of expansion law (dimensionless)
C_v = specific heat capacity at constant volume (kJ/kg K)
C_p = specific heat capacity of constant pressure (kJ/kg K)

Mixtures of gases

$$PV = mR_m T$$

$$m = m_1 + m_2 + m_3$$

$$R_m = \frac{R_1 m_1 + R_2 m_2 + R_3 m_3}{m_1 + m_2 + m_3} = \text{gas constant of mixture}$$

The laws of perfect gases

The **Critical Temperature** of a substance is that temperature above which it cannot exist as a liquid.

The **Critical Pressure** is the pressure of a saturated vapour at its critical temperature.

Critical temperatures and pressures of various substances

Substance	Critical temperature		Critical pressure absolute		Boiling temperature at atmospheric pressure	
	$°F$	$°C$	lb./sq. in.	atm.	$°F$	$°C$
Air	−220	−140	573	39	−	−
Alcohol (C_2H_6O)	421	216	956	65	172.4	78
Ammonia (NH_3)	266	130	1691	115	−27.4	−33
Benzol (C_6H_6)	554	292	735	50	176	80
Carbon-dioxide (CO_2)	88.2	31	1132	77	−110	−79
Carbon-monoxide (CO)	−222	−141	528	35.9	−310	−190
Ether ($C_4H_{10}O$)	381.2	194	544	37	95	35
Hydrogen (H)	−402	−242	294	20	−423	−253
Nitrogen (N)	−236	−149	514	35	−321	−195
Oxygen (O_2)	−180	−118	735	50	−297	−183
Water (H_2O)	706−716	375−380	3200	217.8	212	100

(From Mark's Mech. Eng. Hand.)

Estimations of temperatures of incandescent bodies

Colours of different temperatures

Faint red	960°F	516°C
Dull red	1290°F	700°C
Brilliant red	1470°F	750°C
Cherry red	1650°F	900°C
Bright cherry red	1830°F	1000°C
Orange	2010°F	1100°C
Bright orange	2190°F	1200°C
White heat	2370°F	1300°C
Bright white heat	2550°F	1400°C
Brilliant white heat	2750°F	1500°C

Heat transfer

Transfer of heat may occur by

1 Conduction
2 Convection
3 Radiation

1 Conduction is the transfer of heat through the molecules of a substance.

(a) *Internal Conduction* is transmission within a body.
(b) *External Conduction* is transmission from one body to another, when the two bodies are in contact.

Thermal Conductivity is the heat flowing through one unit of area and one unit of thickness in one unit of time per degree temperature difference.

Thermal Conductance is the heat flowing through a structural component of unit area in unit time per degree temperature difference between its faces.

$$H = \frac{AK(t_2 - t_1)}{X} = AC(t_2 - t_1)$$

$$C = \frac{K}{X}$$

where

H = heat flow, W (Btu/hr)
A = area, m^2 (ft^2)
K = thermal conductivity, W/mK (Btu in/hr ft^2 °F)
C = thermal conductance, W/m^2 K (Btu/hr ft^2 °F)

$X =$ thickness, m (in)
$t_1 =$ temperature at cooler section, °C (°F)
$t_2 =$ temperature at hotter section, °C (°F)

Thermal Resistance is the reciprocal of thermal conductance

$$H = \frac{A(t_2 - t_1)}{R} \qquad \text{W (Btu/hr)}$$
$$R = \frac{1}{C} = \frac{X}{K} \qquad \frac{m^2\,K}{W}\ \frac{\text{hr ft}^2\,°F}{\text{Btu}}$$

2 **Convection** is the transfer of heat by flow of currents within a fluid body. (Liquid or gas flowing over the surface of a hotter or cooler body.)

$$H = aA\,(t_2 - t_1) = \frac{A(t_2 - t_1)}{R_1}\,(\text{Btu/hr or watts})$$

$a =$ Thermal conductance (Btu/hr sq ft °F or W/m^2 °C)

$$R_1 = \frac{1}{aX} = \text{Thermal resistance.}$$

The amount of heat transferred per unit of time is affected by the velocity of moving medium, the area and form of surface and the temperature difference.

3 **Radiation** is the transfer of heat from one body to another by wave motion.

Stephan-Boltzmann Formula

$$E = C\left(\frac{T}{100}\right)^4 \qquad \begin{aligned} &E = \text{Heat emission of a body (Btu/hr or Watts)} \\ &T = \text{Absolute temperature (°R or °K)} \\ &C = \text{Radiation constant} \end{aligned}$$

Quantities of heat transferred between two surfaces:

$$Q_{\text{Rad}} = CA\left[\left(\frac{T_1}{100}\right)^4 - \left(\frac{T_2}{100}\right)^4\right]$$

$A =$ Area

$T_1 T_2 =$ Absolute temperatures of hot and cold surfaces respectively.

For the absolute black body

$C = 5.72$ Watts per sq m per (deg C)^4
$\quad = 0.173$ Btu per hr per sq ft per (deg F)^4

For other materials see table below.

Radiation constant of building material (C)

	$\dfrac{W}{m^2}$ $(°C)^4$	$\dfrac{Btu}{hr\,ft^2}$ $(°F)^4$		$\dfrac{W}{m^2}$ $(°C)^4$	$\dfrac{Btu}{hr\,ft^2}$ $(°F)^4$
Black body	5.72	0.173	Wrought iron, dull oxidised	5.16	0.156
Cotton	4.23	0.128	Wrought iron, polished	1.55	0.047
Glass	5.13	0.155			
Wood	4.17	0.126	Cast iron, rough oxidised	5.09	0.154
Brick	5.16	0.156			
Oil paint	4.30	0.130	Copper, polished	1.19	0.036
Paper	4.43	0.134	Brass, dull	0.152	0.0046
Lamp black	5.16	0.156	Silver	1.19	0.036
Sand	4.20	0.127	Zinc, dull	0.152	0.0046
Shavings	4.10	0.124	Tin	0.26	0.0077
Silk	4.30	0.130	Plaster	5.16	0.156
Water	3.70	0.112			
Wool	4.30	0.130			

Conduction of heat through pipes or partitions

Symbols

t_m = Logarithmic mean temperature difference
t_{a1} = Initial temperature of heating medium
t_{a2} = Final temperature of heating medium
t_1 = Initial temperature of heated fluid
t_2 = Final temperature of heated fluid.

The heat exchange can be classified as follows:

1 Parallel Flow, the fluids flow in the same directions over the separating wall.

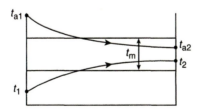

$$t_m = \frac{t_{a1} - t_{a2} + t_2 - t_1}{\log_e \dfrac{(t_{a1} - t_1)}{(t_{a2} - t_2)}} = \frac{\text{Initial temp. dif.} - \text{Final temp. dif.}}{2.3 \log_{10} \dfrac{\text{Initial temp. dif.}}{\text{Final temp. dif.}}}$$

2   Counter Flow, the directions are opposite.

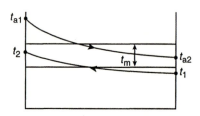

$$t_m = \text{(as before)} = \frac{\text{Initial temp. dif.} - \text{Final temp. dif.}}{2.3 \log_{10} \dfrac{\text{Initial temp. dif.}}{\text{Final temp. dif.}}}$$

3   Evaporators or Condensers
One fluid remains at a constant temperature while changing its state.

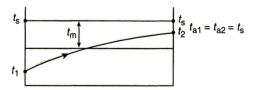

$$t_m = \text{(as before)} = \frac{(t_1 - t_2)}{2.3 \log_{10} \dfrac{t_s - t_2}{t_s - t_1}}$$

4   Mixed Flow
One of the fluids takes an irregular direction with respect to the other.

$$t_m = \frac{t_{a1} - t_{a2}}{2} - \frac{t_1 - t_2}{2}$$

Heat transfer in the unsteady state

Newton's Law of cooling. In the warming and cooling of bodies, the heat gain or loss, respectively, is proportional to the difference between the temperatures of the body and the surroundings.

Let:

θ_s = Temperature of the surroundings °C
θ_1 = Initial temperature of the body °C
θ_2 = Temperature of the body °C
k = Thermal conductivity of the body W/mK
w = Density of the body kg/m^3
s = Specific heat capacity of the body J/kgK
h = Coefficient of heat transfer between the body and the surroundings W/m^{2}K
r = Radius of a sphere or cylinder, or half thickness of a slab cooled or heated on both faces or thickness of a slab cooled or heated on one face only
t = Cooling (or heating) time secs.

Then

$$\frac{\theta_1 - \theta_s}{\theta_2 - \theta_s} = e^{-Kt} \quad \text{and} \quad \log_e(\theta_2 - \theta_s) - \log_e(\theta_1 - \theta_s) = -Kt$$

where K = Constant which can be found by measuring the temperatures of the body at different times t_1 and t_2 and which is given by

$$K = \frac{\log_e(\theta_2 - \theta_s)}{\log_e(\theta_1 - \theta_s)} \Big/ (t_2 - t_1)$$

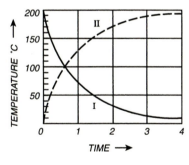
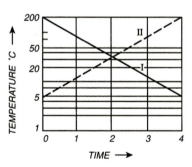

Cooling curve (I) and heating curve (II) showing relation of temperature and time on linear and semi-logarithmic paper.

Graphs showing how the temperature of cooling or heating up bodies can be plotted on semi-logarithmic paper by introducing the following dimensionless ratios

$$Y = \frac{\theta_2 - \theta_s}{\theta_1 - \theta_s} \qquad x = \frac{kt}{wsR^2} \qquad m = \frac{k}{hr}$$

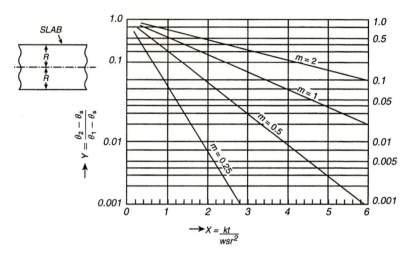

$$\longrightarrow X = \frac{kt}{wsr^2}$$

The Increased Heat Requirement of Buildings during the heating up period causes a greater heat requirement than the steady state. This additional heat loss depends mainly on the type of building, length of heating interruption and heating up time, and type of heating installation. The allowance for covering the increased heat loss during heating up is usually expressed as a percentage of the heat loss in the steady state. See pages 91 and 92.

The temperatures during warming up of bodies are represented graphically by curves which are symmetrical to cooling down curves.

Logarithmic mean temperature differences

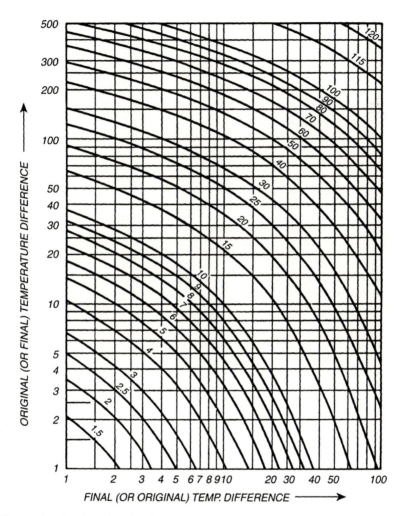

ORIGINAL (OR FINAL) TEMPERATURE DIFFERENCE

FINAL (OR ORIGINAL) TEMP. DIFFERENCE ⟶

Example of using the chart
Water to water calorifier with counter flow
Primary flow temperature 80°C. Secondary return temperature 10°C
Primary return temperature 70°C. Secondary flow temperature 40°C
Original temperature difference = 80−70 = 10°C
Final temperature difference = 70−40 = 30°C
From chart: Log mean temp. dif. = 18°C
The chart can be used equally well for °C or °F.

Transmission of heat

Heat transmission coefficients for metals

			$\dfrac{Watts}{m^2{}^\circ C}$	$\dfrac{Btu}{ft^2 hr{}^\circ F}$
Water	Cast iron	Air or Gas	8.0	1.4
Water	Mild steel	Air or Gas	11.0	2.0
Water	Copper	Air or Gas	11.0	2.25
Water	Cast iron	Water	220 to 280	40 to 50
Water	Mild steel	Water	340 to 400	50 to 70
Water	Copper	Water	350 to 450	62 to 80
Air	Cast iron	Air	6.0	1.0
Air	Mild steel	Air	8.0	1.4
Steam	Cast iron	Air	11.0	2.0
Steam	Mild steel	Air	11.0	2.5
Steam	Copper	Air	17.0	3.0
Steam	Cast iron	Water	900	160
Steam	Mild steel	Water	1050	185
Steam	Copper	Water	1170	205

The above values are average coefficients for practically still fluids. The coefficients are dependent on velocities of heating and heated media – on type of heating surface, temperature difference, and other circumstances. For special cases, see literature and manufacturer's data.

Table of $n^{1.3}$ for radiator and pipe coefficients in relation to various temperature differences

n	$n^{1.3}$	n	$n^{1\,3}$	n	$n^{1.3}$	n	$n^{1.3}$	n	$n^{1.3}$	n	$n^{1.3}$
30	83	70	250	110	450	150	674	190	917	230	1176
35	102	75	273	115	477	155	704	195	948	235	1209
40	121	80	298	120	505	160	733	200	980	240	1242
45	141	85	322	125	533	165	763	205	1012	245	1219
50	162	90	347	130	560	170	793	210	1044	250	1310
55	183	95	372	135	589	175	824	215	1075		
60	205	100	398	140	617	180	855	220	1110		
65	226	105	424	145	645	185	887	225	1142		

Heat loss of steel pipes

For various water temperatures and steam pressures

Nominal bore		Heat loss W/m for fluid inside pipe						Heat loss Btu/h ft for fluid inside pipe					
		Water				Steam		Water				Steam	
in	mm	50°C	60°C	75°C	100°C	1 bar	4 bar	120°F	140°F	170°F	212°F	15 psig	60 psig
½	15	30	40	60	90	130	190	30	40	60	95	135	200
¾	20	35	50	70	110	160	220	35	50	75	115	170	230
1	25	40	60	90	130	190	270	40	60	90	135	200	280
1¼	32	50	70	110	160	230	330	50	70	110	165	240	340
1½	40	55	80	120	180	250	370	55	80	130	190	260	380
2	50	65	95	150	220	310	440	65	90	150	230	320	460
2½	65	80	120	170	260	360	530	80	120	180	270	380	550
3	80	100	140	210	300	440	630	100	140	220	310	460	650
4	100	120	170	260	380	550	800	120	170	270	390	570	830
6	150	170	250	370	540	770	1100	170	250	380	560	800	1150

Correction factors for use with above table

Single pipe along skirting or riser	1.0
More than one pipe along skirting or riser	0.90
Single pipe along ceiling	0.75
More than one pipe along ceiling	0.65
Single pipe freely exposed	1.1
More than one pipe freely exposed	1.0

Heat loss of steel pipes

For high temperature difference (MPHW and HPHW)
For various temperature differences between pipe and air

Nominal bore mm	Heat loss for temperature difference (W/m)							
	110°C	125°C	140°C	150°C	165°C	195°C	225°C	280°C
15	130	155	180	205	235	280	375	575
20	160	190	220	255	290	370	465	660
25	200	235	275	305	355	455	565	815
32	240	290	330	375	435	555	700	1000
40	270	320	375	420	485	625	790	1120
50	330	395	465	520	600	770	975	1390
65	390	465	540	615	715	910	1150	1650
80	470	560	650	740	860	1090	1380	1980
100	585	700	820	925	1065	1370	1740	2520
150	815	970	1130	1290	1470	1910	2430	3500
200	1040	1240	1440	1650	1900	2440	3100	4430
250	1250	1510	1750	1995	2300	2980	3780	5600
300	1470	1760	2060	2340	2690	3370	4430	6450

Nominal bore in	Heat loss for temperature difference (Btu/hr/ft)							
	200°F	225°F	250°F	275°F	300°F	350°F	400°F	500°F
$\frac{1}{2}$	140	160	190	215	245	310	390	595
$\frac{3}{4}$	170	200	230	270	305	380	475	680
1	210	245	285	325	370	470	580	840
$1\frac{1}{4}$	265	300	345	400	460	565	715	1035
$1\frac{1}{2}$	290	335	390	450	510	650	815	1160
2	350	410	480	550	630	800	1000	1460
$2\frac{1}{2}$	430	490	560	650	750	880	1190	1700
3	580	655	670	780	900	1130	1410	2040
4	620	730	850	825	1120	1420	1790	2600
6	860	1010	1170	1360	1540	1980	2700	3600
8	1090	1390	1480	1740	2000	2540	3180	4610
10	1320	1535	1810	2100	2425	3100	3900	5700
12	1530	1835	2125	2460	2830	3500	4950	6650

Heat loss of copper pipes

For various temperature differences between pipe and air

Nominal bore		Heat loss for temperature difference (W/m)			Heat loss for temperature difference (Btu/hr ft)		
in	mm	40°C	55°C	72°C	70°F	100°F	130°F
$\frac{1}{2}$	15	21	32	45	22	34	47
$\frac{3}{4}$	22	28	43	60	29	45	53
1	28	34	53	76	36	56	79
$1\frac{1}{4}$	35	41	64	89	43	67	93
$1\frac{1}{2}$	42	47	74	104	49	77	108
2	54	59	93	131	62	97	136
$2\frac{1}{2}$	67	71	111	156	74	116	162
3	76	83	129	181	87	135	189
4	108	107	165	232	111	172	241

Heat loss of insulated copper pipes

For temperature difference 55°C (100°F)
For 25 mm thick insulation with $k = 0.043$ W/m°C (0.3 Btu in/ft^2 hr °F)

Nominal bore		Heat loss		Nominal bore		Heat loss	
in	mm	W/m	Btu/hr ft	in	mm	W/m	Btu/hr ft
$\frac{3}{4}$	22	8	8	2	54	14.5	15
1	28	10	10	$2\frac{1}{2}$	67	16	17
$1\frac{1}{2}$	42	11.5	12	3	76	19	20

Heat loss through lagging

Insulating material	Heat loss through 75 mm thickness per 55 K difference between faces W/m^2
Asbestos	75
Cork	32
Sawdust	54

Loss for bare metal for 55 K difference is approximately 485 W/m^2

Densities

	kg/m^3	lb/ft^3
Metals		
Aluminium	2690	168
Antimony	6690	417
Brass, cast	8100	505
Bronze, gunmetal	8450	529
Copper	8650	551
Gold, pure, cast	19 200	1200
Iron, cast	7480	467
Iron, wrought	7850	486
Lead	11 340	705
Mercury	13 450	840
Nickel	8830	551
Platinum	21 450	1340
Silver	10 500	655
Steel	7900	493
Tin	7280	455
Zinc	7200	444
Solids		
Asbestos	3060	191
Asphalt	1650	103
Brick	1000–2000	62–134
Cement, Portland	3000	187
Cement, Roman	1550	97
Chalk	1500–2800	95–175
Coal	1500–1650	95–103
Coke	1000	62
Concrete, mean	2240	140
Dowtherm	880–1073	55–67
Glass, window	2640	164
Granite	2130	133
Gypsum	2165	135
Ice, at 0°C	910	57
Lime	2740	171

	kg/m^3	lb/ft^3
Limestone	3170	198
Marble	2650	165
Mortar	1400–1750	86–109
Peat	600–1330	37–83
Plaster	1180	73
Porcelain	2300	143
Rubber	920	67
Salt, common	2130	133
Soap	1070	67
Starch	945	59
Sulphur	2020	126
Wax, paraffin	930	58
Wood	700–900	44–56
Liquids		
Acetic acid	1049	66
Alcohol	790	49
Ammonia	610	38
Beer	1030	64
Ether	870	54
Glycerine	1270	79
Kerosene (paraffin)	810	50
Oil, mineral	850	53
Oil, vegetable	920	57
Milk	1030	64
Petrol	700–750	44–47
Turpentine	870	54
Water, distilled	1000	62
Water, sea, 4°C	1030	64

Specific heat capacities of gases

| Gas | Formula | Specific heat capacity | | | | $\gamma = C_p/C_v$ | Gas constant $= (C_p - C_v)$ | |
| | | Btu/lb °F | | kJ/kg K | | | ft lb/lb °F | kJ/kg K |
		C_p	C_v	C_p	C_v			
Acetylene	C_2H_2	0.350	0.270	1.47	1.13	1.28	59.34	0.34
Air	–	0.251	0.171	1.01	0.716	1.40	53.34	0.29
Ammonia	NH_3	0.523	0.399	2.19	1.67	1.31	96.50	0.52
Blast furnace gas	–	0.245	0.174	1.03	0.729	1.40	55.05	0.297
Carbon dioxide	CO_2	0.210	0.160	0.827	0.632	1.31	38.86	0.189
Carbon monoxide	CO	0.243	0.172	1.02	0.720	1.41	55.14	0.297
Combustion products	–	0.24	–	1.01	–	–	–	–
Ethylene	C_2H_4	0.400	0.330	1.67	1.38	1.20	55.08	0.29
Hydrogen	H_2	3.42	2.44	14.24	10.08	1.40	765.90	4.16
Methane	CH_4	0.593	0.450	2.23	1.71	1.32	111.31	0.60
Nitrogen	N_2	0.247	0.176	1.034	0.737	1.40	54.99	0.297
Oxygen	O_2	0.219	0.157	0.917	0.656	1.40	48.24	0.260
Sulphur dioxide	SO_2	0.154	0.123	0.645	0.515	1.25	24.10	0.130

Density of gases

Gas	Molecular weight	Density at 0°C and atmospheric pressure	
		kg/m^3	lb/ft^3
Acetylene	26	1.170	0.0729
Air	–	1.293	0.0806
Ammonia	17	0.769	0.0480
Blast furnace gas	–	1.250	0.0780
Carbon dioxide	44	1.977	0.1234
Carbon monoxide	28	1.250	0.0780
Combustion products	–	1.11	0.069
Ethylene	28	1.260	0.0786
Hydrogen	2	0.0899	0.0056
Methane	16	0.717	0.0447
Nitrogen	28	1.250	0.0780
Oxygen	32	1.429	0.0892
Sulphur dioxide	64	2.926	0.1828

Specific heat capacities between 0°C and 100°C

	kJ/kg K	Btu/lb °F		kJ/kg K	Btu/lb °F
Metals			Ice	2.11	0.504
Aluminium	0.912	0.218	India rubber	1.1–4.1	0.27–0.98
Antimony	0.214	0.051	Limestone	0.84	0.20
Copper	0.389	0.093	Marble	0.88	0.21
Gold	0.130	0.031	Peat	1.88	0.45
Iron	0.460	0.110	Plaster	0.84	0.20
Lead	0.130	0.031	Porcelain	1.07	0.255
Mercury	0.138	0.033	Sand	0.82	0.19
Nickel	0.452	0.108	Sulphur	0.72	0.17
Platinum	0.134	0.032	Wood	2.3–2.7	0.55–0.65
Silver	0.234	0.056			
Tin	0.230	0.055	**Liquids**		
Zinc	0.393	0.094	Acetic acid	2.13	0.51
			Alcohol	2.93	0.70
Metal alloys			Ammonia	0.47	0.11
Ball metal	0.360	0.086	Benzol	1.80	0.43
Brass	0.377	0.090	Dowtherm	1.55	0.37
Bronze	0.435	0.104	Ether	2.10	0.50
Nickel steel	0.456	0.109	Ethylene glycol	2.38	0.57
Solder	0.167	0.04	Glycerine	2.41	0.58
			Milk	3.93	0.94
Solids			Naphtholene	1.78	0.43
Asbestos	0.84	0.20	Oil, mineral	1.67	0.40
Ashes	0.84	0.20	Oil, vegetable	1.68	0.40
Asphalt	0.80	0.19	Paraffin	2.14	0.51
Brick	0.92	0.22	Petroleum	2.09	0.50
Carbon	0.71	0.17	Sulphuric acid	1.38	0.33
Coke	0.85	0.203	Turpentine	1.98	0.47
Coal	1.31	0.314	Water, fresh	4.19	1.00
Concrete	1.13	0.27	Water, sea	3.94	0.94
Cork	2.03	0.485			
Glass	0.84	0.20			
Granite	0.75	0.18			
Graphite	0.71	0.17			

Boiling points at atmospheric pressure

	$°C$	$°F$		$°C$	$°F$
Alcohol	78	172.4	Hydrogen	−253	−423
Ammonia	−33.4	−28.1	Nitrogen	−196	−320
Aniline	184	363	Oxygen	−183	−297
Carbon dioxide	−78.5	−109.3	Sulphur	440	823
Downtherm	258	496	Toluene	111	230
Ether	35	95	Turpentine	160	320
Glycerine	290	554	Water	100	212
Helium	−269	−452			

Latent heats of vaporisation

	kJ/kg	Btu/lb		kJ/kg	Btu/lb
Alcohol	896	385	Hydrogen	461	198
Ammonia	1369	589	Nitrogen	199	86
Aniline	450	193	Oxygen	214	92
Carbon dioxide	574	247	Sulphur	1510	650
Ether	377	162	Toluene	351	151
Helium	21	9	Turpentine	293	126
			Water	2257	970.4

Melting and solidifying points at atmospheric pressure

	$°C$	$°F$		$°C$	$°F$
Alcohol	−97	−143	Lead	327	621
Aluminium	658	1218	Mercury	−39	−38
Ammonia	−78	−108	Nickel	1455	2646
Aniline	−6	21	Silver	960	1761
Carbon dioxide	−56	−69	Sulphur	106–119	234–247
Copper	1083	1981	Tin	232	449
Dowtherm	12	54	Water	0	32
Glycerine	−16	4	Wax	64	149
Gold	1063	1945	Zinc	419	787
Iron, pure	1530	2786			

Latent heats of melting

	kJ/kg	Btu/lb		kJ/kg	Btu/lb
Aluminium	321	138.2	Lead	22.4	9.65
Ammonia	339	146	Mercury	11.8	5.08
Aniline	113.5	48.8	Nickel	19.4	8.35
Carbon dioxide	184	79	Silver	88.0	37.9
Copper	176	75.6	Sulphur	39.2	16.87
Glycerine	176	75.6	Tin	58.5	25.2
Iron, grey cast	96	41.4	Water	334	144
Iron, white cast	138	59.4	Zinc	118	50.63
Iron slag	209	90.0			

Coefficients of linear expansion Average values between 0°C and 100°C

	$\dfrac{m}{mK} \times 10^6$	$\dfrac{in}{in\,°F} \times 10^6$		$\dfrac{m}{mK} \times 10^6$	$\dfrac{in}{in\,°F} \times 10^6$
Aluminium	22.2	12.3	Lead	28.0	15.1
Antimony	10.4	5.8	Marble	12	6.5
Brass	18.7	10.4	Masonry	4.5–9.0	2.5–9.0
Brick	5.5	3.1	Mortar	7.3–13.5	4.1–7.5
Bronze	18.0	10.0	Nickel	13.0	7.2
Cement	10.0	6.0	Plaster	25	13.9
Concrete	14.5	8.0	Porcelain	3.0	1.7
Copper	16.5	9.3	Rubber	77	42.8
Glass, hard	5.9	3.3	Silver	19.5	10.7
Glass, plate	9.0	5.0	Solder	24.0	13.4
Gold	14.2	8.2	Steel, nickel	13.0	7.3
Graphite	7.9	4.4	Type metal	19.0	10.8
Iron, pure	12.0	6.7	Wood, oak parallel to grain	4.9	2.7
Iron, cast	10.4	5.9	Wood, oak across grain	5.4	3.0
Iron, forged	11.3	6.3	Zinc	29.7	16.5

Thermal properties of water

Temp. °F	Abs. pressure lb/in²	Density lb/ft³	Specific gravity	Specific volume ft³/lb	Specific heat Btu lb °F	Specific entropy Btu lb °F	Dynamic viscosity in poises	Specific enthalpy Btu/lb
32	0.088	62.42	1.000	0.0160	1.0093	0.0000	0.0179	0
40	0.122	62.42	1.000	0.0160	1.0048	0.01615	0.0155	8
50	0.178	62.42	1.000	0.0160	1.0015	0.03595	0.0131	18
60	0.256	62.38	1.000	0.0160	0.9995	0.05765	0.0113	28
62	0.275	62.35	1.000	0.0160	0.9992	0.05919	0.0110	30
70	0.363	62.30	0.999	0.0160	0.9982	0.0754	0.0098	38
80	0.507	62.22	0.998	0.0160	0.9975	0.0929	0.0086	48
90	0.698	62.11	0.996	0.0161	0.9971	0.1112	0.0076	58
100	0.949	61.99	0.994	0.0161	0.9970	0.1292	0.0068	68
110	1.27	61.86	0.992	0.0161	0.9971	0.1469	0.0062	78
120	1.69	61.71	0.990	0.0162	0.9974	0.1641	0.0056	88
130	2.22	61.55	0.987	0.0162	0.9978	0.1816	0.0051	98
140	2.89	61.38	0.984	0.0163	0.9984	0.1981	0.0047	108
150	3.72	61.20	0.982	0.0163	0.9990	0.2147	0.0043	118
160	4.74	61.00	0.979	0.0164	0.9988	0.2309	0.0040	128
170	5.99	60.80	0.975	0.0164	1.0007	0.2472	0.0037	138
180	7.51	60.58	0.971	0.0165	1.0017	0.2629	0.00345	148
190	9.33	60.36	0.969	0.0166	1.0028	0.2787	0.00323	158
200	11.53	60.12	0.965	0.0166	1.0039	0.2938	0.00302	168
210	14.13	59.92	0.958	0.0167	1.0052	0.3089	0.00287	178
212	14.70	59.88	0.957	0.0167	1.0055	0.3118	0.00285	180
220	17.19	59.66	0.955	0.0168	1.0068	0.3237	0.00272	188.1
230	20.77	59.37	0.950	0.0168	1.0087	0.3385	0.00257	198.2
240	24.97	59.17	0.946	0.0169	1.0104	0.3531	0.00254	208.3
250	29.81	58.84	0.941	0.0170	1.0126	0.3676	0.00230	218.4
260	35.42	58.62	0.940	0.0171	1.0148	0.3818	0.00217	228.6
270	41.85	58.25	0.933	0.0172	1.0174	0.3962	0.00208	238.7
280	49.18	58.04	0.929	0.0172	1.0200	0.4097	0.00200	248.9
290	57.55	57.65	0.923	0.0173	1.0230	0.4236	0.00193	259.2
300	67.00	57.41	0.920	0.0174	1.0260	0.4272	0.00186	262.5
310	77.67	57.00	0.913	0.0175	1.0296	0.4507	0.00179	279.8
320	89.63	56.65	0.906	0.0177	1.0332	0.4643	0.00173	290.2
330	103.00	56.31	0.900	0.0178	1.0368	0.4777	0.00168	300.6
340	118.0	55.95	0.897	0.0179	1.0404	0.4908	0.00163	311.1
350	134.6	55.65	0.890	0.0180	1.0440	0.5040	0.00158	321.7
360	153.0	55.19	0.883	0.0181	1.0486	0.5158	0.00153	332.3
370	173.3	54.78	0.876	0.0182	1.0532	0.5292	0.00149	342.9
380	195.6	54.36	0.870	0.0184	1.0578	0.5420	0.00145	353.5
390	220.2	53.96	0.865	0.0187	1.0624	0.5548	0.00141	364.3
400	247.1	53.62	0.834	0.0186	1.0670	0.5677	0.00137	375.3
450	422	51.3	0.821	0.0195	1.0950	0.6298	–	430.2
500	679	48.8	0.781	0.0205	1.1300	0.6907	–	489.1
550	1043	45.7	0.730	0.0219	1.2000	0.7550	–	553.5
600	1540	41.5	0.666	0.0241	1.3620	0.8199	–	623.2
706.1	3226	19.2	0.307	0.0522	–	1.0785	–	925.0

Thermal properties of water

Temp °C	Abs pressure kN/m²	Density kg/m³	Specific volume m³/kg	Specific heat capacity kJ/kg K	Specific entropy kJ/kg K	Dynamic viscosity centipoise	Specific enthalpy kJ/kg
0	0.6	1000	0.00100	4.217	0	1.78	0
5	0.9	1000	0.00100	4.204	0.075	1.52	21.0
10	1.2	1000	0.00100	4.193	0.150	1.31	41.9
15	1.7	999	0.00100	4.186	0.223	1.14	62.9
20	2.3	990	0.00100	4.182	0.296	1.00	83.8
25	3.2	997	0.00100	4.181	0.367	0.890	104.8
30	4.3	996	0.00100	4.179	0.438	0.798	125.7
35	5.6	994	0.00101	4.178	0.505	0.719	146.7
40	7.7	991	0.00101	4.179	0.581	0.653	167.6
45	9.6	990	0.00101	4.181	0.637	0.596	188.6
50	12.5	988	0.00101	4.182	0.707	0.547	209.6
55	15.7	986	0.00101	4.183	0.767	0.504	230.5
60	20.0	980	0.00102	4.185	0.832	0.467	251.5
65	25.0	979	0.00102	4.188	0.893	0.434	272.4
70	31.3	978	0.00102	4.190	0.966	0.404	293.4
75	38.6	975	0.00103	4.194	1.016	0.378	314.3
80	47.5	971	0.00103	4.197	1.076	0.355	335.3
85	57.8	969	0.00103	4.203	1.134	0.334	356.2
90	70.0	962	0.00104	4.205	1.192	0.314	377.2
95	84.5	962	0.00104	4.213	1.250	0.297	398.1
100	101.33	962	0.00104	4.216	1.307	0.281	419.1
105	121	955	0.00105	4.226	1.382	0.267	440.2
110	143	951	0.00105	4.233	1.418	0.253	461.3
115	169	947	0.00106	4.240	1.473	0.241	482.5
120	199	943	0.00106	4.240	1.527	0.230	503.7
125	228	939	0.00106	4.254	1.565	0.221	524.3
130	270	935	0.00107	4.270	1.635	0.212	546.3
135	313	931	0.00107	4.280	1.687	0.204	567.7
140	361	926	0.00108	4.290	1.739	0.196	588.7
145	416	922	0.00108	4.300	1.790	0.190	610.0
150	477	918	0.00109	4.310	1.842	0.185	631.8
155	543	912	0.00110	4.335	1.892	0.180	653.8
160	618	907	0.00110	4.350	1.942	0.174	674.5
165	701	902	0.00111	4.364	1.992	0.169	697.3
170	792	897	0.00111	4.380	2.041	0.163	718.1
175	890	893	0.00112	4.389	2.090	0.158	739.8
180	1000	887	0.00113	4.420	2.138	0.153	763.1
185	1120	882	0.00113	4.444	2.187	0.149	785.3
190	1260	876	0.00114	4.460	2.236	0.145	807.5
195	1400	870	0.00115	4.404	2.282	0.141	829.9
200	1550	863	0.00116	4.497	2.329	0.138	851.7
225	2550	834	0.00120	4.648	2.569	0.121	966.8
250	3990	800	0.00125	4.867	2.797	0.110	1087
275	5950	756	0.00132	5.202	3.022	0.0972	1211
300	8600	714	0.00140	5.769	3.256	0.0897	1345
325	12 130	654	0.00153	6.861	3.501	0.0790	1494
350	16 540	575	0.00174	10.10	3.781	0.0648	1672
360	18 680	526	0.00190	14.60	3.921	0.0582	1764

Properties of water

Density: At 4 °C 1 litre = 1 kg
At 62°F 1 gal = 10 lb

Freezing temperature	0°C	32°F
Boiling temperature	100°C	212°F
Latent heat of melting	334 kJ/kg	144 Btu/lb
Latent heat of evaporation	2,270 kJ/kg	977 Btu/lb
Critical temperature	380–386°C	706–716°F
Critical pressure	23,520 kN/m^2	3,200 lb/in^2
Specific heat capacity		
water	4.187 kJ/kg K	1.00 Btu/lb °F
ice	2.108 kJ/kg K	0.504 Btu/lb °F
water vapour	1.996 kJ/kg K	0.477 Btu/lb °F

Thermal expansion

From 4°C to 100°C water expands by $\frac{1}{24}$ of its original volume.

Bulk modulus of elasticity	2,068,500 kN/m^2
	300,000 lb/in^2

5 Properties of steam and air

Properties of steam and other vapours

A **vapour** is any substance in the gaseous state which does not even approximately follow the general gas laws.

Highly superheated vapours are gases, if the superheat is sufficiently great, and they then approximately follow the general gas law.

Conditions of vapours

1 **Dry Saturated vapour** is free from unvaporised liquid particles.
2 **Wet Saturated vapour** carries liquid globules in suspension.
3 **Superheated vapour** is vapour the temperature of which is higher than that of the boiling point corresponding to the pressure.

Dryness fraction or quality of saturated vapour (X) is the percentage of dry vapour present in the given amount of the wet saturated vapour.

$$X = \frac{W_s}{W_s + W_w} \times 100\%$$

W_s = Weight of dry steam in steam considered
W_w = Weight of water in steam

The heat of the liquid 'h' is the heat in Joules per kg required to raise the temperature of the liquid from 0°C to the temperature at which the liquid begins to boil at the given pressure.

$$h = ct$$

c = Mean specific heat capacity of water
t = Temperature of formation of steam at pressure considered °C

The latent heat of evaporation 'L' is the heat required to change a liquid at a given temperature and pressure into a vapour at the same temperature and pressure. It is divided into two parts

1 External latent heat of vapour = External work heat
2 Internal latent heat of vapour = Heat due to change of state

The total heat of a vapour (or enthalpy) is the amount of heat which must be supplied to 1 kg of the liquid which is at 0°C to convert it at constant pressure into vapour at the temperature and pressure considered.

Total heat of dry saturated vapour

$$H = h + L \text{ (Joules per kg)}$$

h = Heat of liquid at the temperature of the wet vapour, Joules per kg
L = Latent heat, Joules per kg

Total heat of wet saturated vapour

$$H_w = h + xL \text{ (Joules per kg)}$$

$x =$ Dryness factor

Total heat of superheated vapour

$$H_s = h + L + c(t_s - t) \text{ (Joules per kg)}$$

$c =$ Mean specific heat capacity of superheated vapour at the pressure and degree of superheat considered
$t_s =$ Temperature of superheat °C
$t_1 =$ Temperature of formation of steam °C

Specific volumes of wet vapour

$$V_w = (1 - x)V + xV_D$$

$$V_w = xV_D, \quad x = \frac{V_w}{V_D}, \quad \text{when } x = \text{very small}$$

$V_w =$ Specific volume of the wet vapour, m^3 per kg
$V_D =$ Specific volume of dry saturated vapour of the same pressure, m^3 per kg (Can be found from the Vapour Tables).

Specific volume of superheated vapour
Approximate method by using Charles' Law

$$V = \frac{V_s T_s}{T_1}$$

Entropy of steam

1 **Entropy of water**
Change of Entropy $= \log_e (T_1/T)$
$T_1, T =$ Absolute temperature.
Entropy of water above freezing point $= \phi_w = \log_e (T_1/273)$

2 **Entropy of evaporation**
Change of Entropy during evaporation $= dL/T$
Entropy of 1 kg of wet steam above freezing point

$$\phi_s = \phi_w + \frac{xL_1}{T_1}$$

3 **Entropy of superheated steam**
Change of entropy per kg of steam during superheating $= C_p \log_e (T/T_1)$

Total entropy of 1 kg of superheated steam above freezing point

$$= \phi_w + \frac{L_1}{T_1} + C_p \log_e \frac{T_s}{T_1}$$

$L_1 =$ Latent heat of evaporation at $T_1°C$ absolute
$T_1 =$ Absolute temperature of evaporation
$T_s =$ Absolute temperature of superheat

Temperature − entropy diagram for steam

Shows the relationship between Pressure, Temperature, Dryness Fraction and Entropy.
   When two of these factors are given the two others can be found on the chart.
   The ordinates represent the Absolute Temperature and the Entropy.
   The chart consists of the following lines:

1   Isothermal lines
2   Pressure lines
3   Lines of dryness fraction
4   Water line between water and steam
5   Dry steam lines
6   Constant volume lines

   The total heat is given by the area, enclosed by absolute zero base water line and horizontal and vertical line from the respective points.

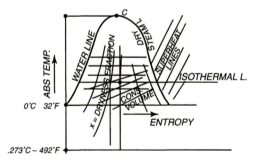

An adiabatic expansion is a vertical line (expansion at constant entropy, no transfer of heat).
$C =$ Critical temperature of steam
   $= 706°F$ to $716°F$
   $= 375°C$ to $380°C$

Critical pressure: $3200 \text{ lb/in}^2 = 217.8 \text{ atm} = 23,500 \text{ kN/m}^2$

Mollier or total heat − entropy chart

Contains the same lines as the temperature–entropy diagram, but with ordinates representing the total heat and entropy of steam. This diagram is used to find the drop in the total heat of steam during an adiabatic expansion.

Total heat of superheated steam (Btu per lb)

Abs. Pres. lb/in^2	Sat. temp. °F	Degrees of Superheat °F						
		0	40	80	120	160	200	280
20	228	1157.1	1177.2	1197.2	1216.9	1236.6	1256.1	1294.9
30	250.3	1165.5	1185.9	1206.1	1226.1	1245.9	1265.6	1304.7
40	267.2	1171.6	1192.3	1212.9	1233.0	1253.0	1272.8	1312.2
50	280.9	1176.3	1197.3	1218.1	1238.5	1258.6	1278.5	1317.9
60	292.6	1180.1	1201.4	1222.2	1242.8	1263.1	1283.2	1322.9
70	302.8	1183.3	1204.7	1225.8	1246.6	1266.9	1287.2	1327.1
80	311.9	1186.1	1207.9	1229.1	1250.0	1270.5	1290.9	1330.8
90	320.2	1188.5	1210.5	1232.1	1253.0	1273.7	1294.0	1334.2
100	327.9	1190.7	1212.9	1234.6	1255.7	1276.5	1297.0	1337.3
120	341.3	1194.3	1216.9	1239.0	1260.4	1281.3	1302.0	1342.6
140	353.0	1197.2	1220.2	1242.5	1264.2	1285.5	1306.3	1347.1
160	363.6	1199.7	1222.9	1245.6	1267.6	1289.1	1310.0	1351.1
180	373.1	1201.7	1225.5	1248.3	1270.7	1292.2	1313.2	1354.6
200	381.8	1203.5	1227.6	1250.7	1273.1	1295.0	1316.2	1358.0
250	401.0	1207.0	1231.7	1255.7	1278.9	1301.2	1322.6	1364.9
300	417.4	1209.4	1235.0	1259.5	1283.2	1305.8	1327.6	1370.3
400	444.7	1212.1	1239.6	1265.4	1289.9	1313.3	1335.8	1379.6
500	467.1	1213.2	1242.2	1269.1	1294.7	1318.8	1341.9	1386.6

Entropy of superheated steam (Btu per °F per lb)

Abs. Pres. lb/in^2	Sat. temp. °F	Degrees of Superheat °F						
		0	40	80	120	160	200	280
20	228	1.7333	1.7617	1.7883	1.8134	1.8372	1.8596	1.9017
30	250.3	1.7017	1.7298	1.7560	1.7807	1.8041	1.8261	1.8472
40	267.2	1.6793	1.7071	1.7331	1.7575	1.7806	1.8025	1.8233
50	280.9	1.6619	1.6895	1.7153	1.7397	1.7626	1.7843	1.8049
60	292.6	1.6477	1.6752	1.7010	1.7253	1.7480	1.7694	1.7899
70	302.8	1.6357	1.6632	1.6889	1.7130	1.7357	1.7570	1.7774
80	311.9	1.6254	1.6527	1.6784	1.7024	1.7251	1.7463	1.7665
90	320.2	1.6161	1.6436	1.6692	1.6931	1.7157	1.7367	1.7569
100	327.9	1.6079	1.6353	1.6608	1.6847	1.7073	1.7283	1.7484
120	341.3	1.5935	1.6210	1.6467	1.6705	1.6928	1.7138	1.7337
140	353.0	1.5813	1.6088	1.6345	1.6583	1.6805	1.7014	1.7212
160	363.6	1.5706	1.5983	1.6240	1.6479	1.6701	1.6909	1.7107
180	373.1	1.5610	1.5890	1.6148	1.6386	1.6607	1.6815	1.7013
200	381.8	1.5525	1.5806	1.6063	1.6301	1.6523	1.6730	1.6929
250	401.0	1.5342	1.5628	1.5886	1.6125	1.6347	1.6554	1.6751
300	417.4	1.5190	1.5479	1.5740	1.5980	1.6203	1.6410	1.6607
400	444.7	1.4941	1.5240	1.5506	1.5749	1.5973	1.6181	1.6379
500	467.1	1.4740	1.5049	1.5322	1.5568	1.5795	1.6004	1.6201

Properties of saturated steam

(Based on Callendar's Values)

Abs. pres. p lb in²	Temp. t °F	Specific volume v ft³/lb	Density w lb/ft³	Heat of Liquid h Btu/lb	Evap. L Btu/lb	Sat. vap. H Btu/lb	Entropy S Btu/lb °F
0.5	79.5	640.5	0.00156	47.4	1045	1092	2.0299
1	101.7	333.1	0.0030	69.5	1033	1102	1.9724
2	126.1	173.5	0.0058	93.9	1020	1114	1.9159
3	141.5	118.6	0.0085	109.3	1012	1121	1.8833
4	153.0	90.5	0.0111	120.8	1005	1126	1.8600
5	162.3	73.4	0.0136	130.1	1000	1130	1.8422
6	170.1	61.9	0.0162	137.9	995	1133	1.8277
7	176.9	53.6	0.0187	144.8	991	1136	1.8156
8	182.9	47.3	0.0212	150.8	988	1139	1.8049
9	188.3	42.4	0.0236	156.3	985	1141	1.7956
10	193.2	38.4	0.0261	161.1	982	1143	1.7874
12	202.0	32.4	0.0309	169.9	977	1147	1.7731
14	209.6	28.0	0.0357	177.6	972	1150	1.7611
14.7	212.0	26.8	0.0373	180.0	970	1151	1.7573
16	216.3	24.7	0.0404	184.4	968	1152.5	1.7506
18	222.4	22.2	0.0451	190.5	964	1155	1.7414
20	228.0	20.1	0.0498	196.1	961	1157	1.7333
22	233.1	18.37	0.0545	201.3	958	1159	1.7258
24	237.8	16.93	0.0591	206.1	955	1161	1.7189
26	242.2	15.71	0.0636	210.5	952	1162.5	1.7126
28	246.4	14.66	0.0682	214.8	949	1164	1.7069
30	250.3	13.72	0.0728	218.8	947	1165.5	1.7017
35	259.3	11.86	0.0843	228	941	1169	1.6898
40	267.2	10.48	0.0953	236	936	1172	1.6793
45	274.4	9.37	0.1067	243	931	1174	1.6701
50	281.0	8.50	0.1175	250	926	1176	1.6619
55	287.0	7.74	0.1292	256	922	1178	1.6547
60	292.6	7.16	0.1397	262	919	1180	1.6479
65	297.9	6.64	0.1506	267	914	1182	1.6415
70	303.0	6.20	0.1613	272	911	1183	1.6357
75	307.5	5.81	0.1721	277	907	1185	1.6304
80	312.0	5.47	0.1828	282	904	1186	1.6254
85	316.2	5.16	0.1938	286	901	1187	1.6206
90	320.2	4.89	0.2045	290	898	1189	1.6161
95	324.1	4.65	0.2150	295	895	1190	1.6120
100	327.9	4.43	0.2257	298	893	1191	1.6079
105	331.4	4.23	0.2364	302	890	1192	1.6041
110	334.8	4.04	0.2475	306	887	1193	1.6004
115	338.1	3.88	0.2577	309	884	1194	1.5969

Properties of saturated steam

Abs. pres. p lb in^2	Temp. t °F	Specific volume v ft^3/lb	Density w lb/ft^3	Heat of Liquid h Btu/lb	Evap. L Btu/lb	Sat. vap. H Btu/lb	Entropy S Btu/lb °F
120	341.3	3.73	0.2681	312	882	1194	1.5935
125	344.4	3.59	0.2786	316	879	1195	1.5903
130	347.3	3.46	0.2890	319	877	1196	1.5872
135	350.2	3.33	0.3003	322	875	1197	1.5842
140	353.0	3.22	0.3106	325	872	1197	1.5813
145	355.8	3.12	0.3205	328	870	1198	1.5785
150	358.4	3.02	0.3311	331	868	1199	1.5758
160	363.6	2.84	0.3521	336	864	1200	1.5706
170	368.4	2.68	0.3731	341	860	1201	1.5657
180	373.1	2.54	0.3937	346	856	1202	1.5610
190	377.5	2.41	0.4149	351	852	1203	1.5567
200	382	2.29	0.4347	356	848	1203	1.5525
220	390	2.09	0.4785	364	841	1205	1.5448
240	387	1.93	0.5181	372	834	1206	1.5376
260	404.5	1.78	0.5618	380	827	1207.5	1.5310
280	411.1	1.66	0.6024	387	821	1208.5	1.5241
300	417.4	1.55	0.6452	394	815	1209.4	1.5190
350	431.8	1.34	0.7463	410	801	1211.1	1.5058
400	444.7	1.17	0.8547	425	787.5	1212.1	1.4941
450	456.4	1.04	0.9615	437.8	775	1212.8	1.4836
500	467.1	0.94	1.0638	450.1	763.1	1213.2	1.4740

Specific enthalpy of superheated steam (kJ/kg)

Absolute pressure kN/m^2	Sat. temp. $°C$	Steam temperature $°C$						
		120	150	180	200	230	250	280
150	114.4	2711	2772	2832	2872	2932	2972	3033
200	120.2	–	2769	2830	2870	2931	2971	3031
250	127.4	–	2765	2827	2868	2929	2969	3030
350	138.9	–	2756	2821	2863	2925	2966	3028
400	143.6	–	2752	2818	2860	2923	2964	3026
500	151.8	–	–	2811	2855	2919	2961	3023
600	158.8	–	–	2805	2850	2915	2958	3021
700	165.0	–	–	2798	2844	2911	2954	3018
800	170.4	–	–	2791	2839	2907	2950	3015
900	175.4	–	–	2784	2833	2902	2947	3012
1000	179.9	–	–	2777	2827	2898	2943	3009
1100	184.1	–	–	–	2821	2893	2939	3006
1200	188.0	–	–	–	2814	2889	2935	3003
1400	195.0	–	–	–	2801	2879	2928	2997
1600	201.4	–	–	–	–	2869	2919	2991
2000	212.4	–	–	–	–	2848	2902	2978
2500	223.9	–	–	–	–	2820	2880	2960
3500	242.5	–	–	–	–	–	2828	2922

Specific entropy of superheated steam (kJ/kg K)

Absolute pressure kN/m^2	Sat. temp. $°C$	Steam temperature $°C$						
		120	150	180	200	230	250	280
150	111.4	7.269	7.419	7.557	7.664	7.767	7.845	7.957
200	120.2	–	7.279	7.420	7.507	7.631	7.710	7.822
250	127.4	–	7.169	7.311	7.400	7.525	7.604	7.717
350	138.9	–	6.998	7.146	7.237	7.364	7.444	7.558
400	143.6	–	6.929	7.079	7.171	7.299	7.380	7.495
500	151.8	–	–	6.965	7.059	7.190	7.272	7.388
600	158.8	–	–	6.869	6.966	7.100	7.183	7.300
700	165.0	–	–	6.786	6.886	7.022	7.107	7.225
800	170.4	–	–	6.712	6.815	6.954	7.040	7.156
900	175.4	–	–	6.645	6.751	6.893	6.980	7.101
1000	179.9	–	–	6.584	6.692	6.838	6.926	7.049
1100	184.1	–	–	–	6.638	6.787	6.876	7.001
1200	188.0	–	–	–	6.587	6.739	6.831	6.956
1400	195.0	–	–	–	6.494	6.653	6.748	6.877
1600	201.4	–	–	–	–	6.577	6.674	6.806
2000	212.4	–	–	–	–	6.440	6.546	6.685
2500	223.9	–	–	–	–	6.292	6.407	6.558
3500	242.5	–	–	–	–	–	6.173	6.349

Properties of saturated steam

Absolute pressure kN/m^2	Temp. $°C$	Specific volume m^3/kg	Density kg/m^3	Specific enthalpy of			Specific entropy of steam $kJ/kg\ K$
				Liquid kJ/kg	Evaporation kJ/kg	Steam kJ/kg	
0.8	3.8	160	0.00626	15.8	2493	2509	9.058
2.0	17.5	67.0	0.0149	73.5	2460	2534	8.725
5.0	32.9	28.2	0.0354	137.8	2424	2562	8.396
10.0	45.8	14.7	0.0682	191.8	2393	2585	8.151
20.0	60.1	7.65	0.131	251.5	2358	2610	7.909
28	67.5	5.58	0.179	282.7	2340	2623	7.793
35	72.7	4.53	0.221	304.3	2327	2632	7.717
45	78.7	3.58	0.279	329.6	2312	2642	7.631
55	83.7	2.96	0.338	350.6	2299	2650	7.562
65	88.0	2.53	0.395	368.6	2288	2657	7.506
75	91.8	2.22	0.450	384.5	2279	2663	7.457
85	95.2	1.97	0.507	398.6	2270	2668	7.415
95	98.2	1.78	0.563	411.5	2262	2673	7.377
100	99.6	1.69	0.590	417.5	2258	2675	7.360
101.33	100	1.67	0.598	419.1	2257	2676	7.355
110	102.3	1.55	0.646	428.8	2251	2680	7.328
130	107.1	1.33	0.755	449.2	2238	2687	7.271
150	111.4	1.16	0.863	467.1	2226	2693	7.223
170	115.2	1.03	0.970	483.2	2216	2699	7.181
190	118.6	0.929	1.08	497.8	2206	2704	7.144
220	123.3	0.810	1.23	517.6	2193	2711	7.095
260	128.7	0.693	1.44	540.9	2177	2718	7.039
280	131.2	0.646	1.55	551.4	2170	2722	7.014
320	135.8	0.570	1.75	570.9	2157	2728	6.969
360	139.9	0.510	1.96	588.5	2144	2733	6.930
400	143.6	0.462	2.16	604.7	2133	2738	6.894
440	147.1	0.423	2.36	619.6	2122	2742	6.862
480	150.3	0.389	2.57	633.5	2112	2746	6.833
500	151.8	0.375	2.67	640.1	2107	2748	6.819
550	155.5	0.342	2.92	655.8	2096	2752	6.787
600	158.8	0.315	3.175	670.4	2085	2756	6.758
650	162.0	0.292	3.425	684.1	2075	2759	6.730
700	165.0	0.273	3.66	697.1	2065	2762	6.705
750	167.8	0.255	3.915	709.3	2056	2765	6.682
800	170.4	0.240	4.16	720.9	2047	2768	6.660
850	172.9	0.229	4.41	732.0	2038	2770	6.639
900	175.4	0.215	4.65	742.6	2030	2772	6.619
950	177.7	0.204	4.90	752.8	2021	2774	6.601
1000	179.9	0.194	5.15	762.6	2014	2776	6.583

Properties of saturated steam (continued)

Absolute pressure kN/m²	Temp. °C	Specific volume m³/kg	Density kg/m³	Specific enthalpy of Liquid kJ/kg	Evaporation kJ/kg	Steam kJ/kg	Specific entropy of steam kJ/kg K
1050	182.0	0.186	5.39	772	2006	2778	6.566
1150	186.0	0.170	5.89	790	1991	2781	6.534
1250	189.8	0.157	6.38	807	1977	2784	6.505
1300	191.6	0.151	6.62	815	1971	2785	6.491
1500	198.3	0.132	7.59	845	1945	2790	6.441
1600	201.4	0.124	8.03	859	1933	2792	6.418
1800	207.1	0.110	9.07	885	1910	2795	6.375
2000	212.4	0.0995	10.01	909	1889	2797	6.337
2100	214.9	0.0949	10.54	920	1878	2798	6.319
2300	219.6	0.0868	11.52	942	1858	2800	6.285
2400	221.8	0.0832	12.02	952	1849	2800	6.269
2600	226.0	0.0769	13.01	972	1830	2801	6.239
2700	228.1	0.0740	13.52	981	1821	2802	6.224
2900	232.0	0.0689	14.52	1000	1803	2802	6.197
3000	233.8	0.0666	15.00	1008	1794	2802	6.184
3200	237.4	0.0624	16.02	1025	1779	2802	6.158
3400	240.9	0.0587	17.04	1042	1760	2802	6.134
3600	244.2	0.0554	18.06	1058	1744	2802	6.112
3800	247.3	0.0524	19.08	1073	1728	2801	6.090
4000	250.3	0.0497	21.0	1087	1713	2800	6.069

Taken by permission of Cambridge University Press from *Thermodynamic Tables in S.I. (metric) Units* by Haywood.

Properties of air

Symbols

V	volume of air-vapour mixture	m^3
m	mass of air-vapour mixture	kg
p_a	partial pressure of dry air	N/m^2
p_{wa}	actual partial pressure of water vapour	N/m^2
p_{ws}	saturation pressure of water vapour	N/m^2
p_t	total pressure of mixture	N/m^2
t	dry bulb temperature	°C
T	absolute dry bulb temperature $= 273 + t$	K
ϕ	relative humidity	per cent
X	specific humidity of air-vapour mixture	g/kg
X_s	specific humidity of saturated air	g/kg
ρ_a	density of dry air	kg/m^3
ρ_w	density of water vapour	kg/m^3
ρ	density of air-vapour mixture	kg/m^3
R	gas constant	J/kg K
	$= 286$ for air	
	$= 455$ for water vapour	

Atmospheric air is a mixture of dry air and water vapour. It can be treated as an ideal gas without great discrepancies and the gas laws can be applied to it.

General Gas Law
$$pV = mRT$$

$$\rho = \frac{m}{V} = \frac{p}{RT}$$

Density of Dry Air
$$\rho_a = 0.00350 \frac{p_a}{T}$$

Density of Water Vapour
$$\rho_w = 0.00220 \frac{p_w}{T}$$

Density of Air-Water Vapour Mixture
$$\rho = 0.00350 \frac{p_t}{T} - 0.00133 \frac{p_{ws}\phi}{100T}$$

Air-water vapour mixture is always lighter than dry air.

Humidity is the term applied to the quantity of water vapour present in the air.

Absolute Humidity is the actual mass of water vapour present, expressed in grams water vapour per kilogram mixture.

Specific Humidity is the actual mass of water vapour present, expressed in grams water vapour per kilogram dry air.

$$X = \frac{622\phi\rho_{ws}}{(\rho - \rho_{ws})100} \text{ g/kg}$$

Specific Humidity of Saturated Air

$$X_s = \frac{622\rho_{ws}}{\rho - \rho_{ws}} \text{ g/kg}$$

Relative Humidity is

either   ratio of actual partial pressure of water vapour to vapour pressure at saturation at actual dry bulb temperature.

or   ratio of actual vapour density to vapour density at saturation at actual dry bulb temperature.

or   ratio of actual mass of water vapour in given air volume to mass of water vapour required to saturate this volume.

It is usually expressed in %

$$\phi = \frac{p_{wa}}{p_{ws}} \times 100 = \frac{\rho_w}{\rho_{ws}} \times 100 = \frac{X}{X_s} \times 100\%$$

Saturated Air holds the maximum mass of water vapour at the given temperature. Any lowering of the air temperature will cause condensation of water vapour.

Dry Bulb Temperature is the air temperature as indicated by a thermometer which is not affected by the moisture of the air.

Wet Bulb Temperature is the temperature of adiabatic saturation. It is the temperature indicated by a moistened thermometer bulb exposed to a current of air.

Dew Point Temperature is the temperature to which air with a given moisture content must be cooled to produce saturation of the air and the commencement of condensation of the vapour in the air.

Specific Enthalpy of dry air

$$H = 1.01t \, \text{kJ/kg}$$

Specific Enthalpy of air-water vapour mixture is composed of the sensible heat of the air and the latent heat of vaporisation of the water vapour present in the air and the sensible heat of the vapour.

$$H = 1.01t + X(2463 + 1.88t) \, \text{kJ/kg}$$

1.01 is the specific heat capacity of dry air.
2463 is the latent heat of vaporisation of water at 0°C.
1.88 is the specific heat capacity of water vapour at constant pressure.

Thermal expansion of air
Dry air expands or contracts uniformly 1/886 of its volume per °C under constant pressure.

Humidity Chart for Air (Psychrometric Chart)
See Chart No. 5. The chart shows the relationship between

1 Dry bulb temperature.
2 Wet bulb temperature.
3 Dew point.
4 Relative humidity.
5 Moisture content.
6 Specific volume.
7 Specific enthalpy.

When any two of these are given the other five can be read from the chart. The chart contains the following lines

   (i)   Lines of constant temperature.
  (ii)   Lines of constant specific enthalpy.
 (iii)   Lines of constant wet bulb temperature.
  (iv)   Lines of constant relative humidity.
   (v)   Lines of constant dewpoint, which are also lines of constant moisture content.
  (vi)   Lines of constant specific volume.

Specific Heat Capacity of dry air

$$s = 1.01 \, \text{kJ/kg K}$$

$$= 1.23 \, \text{kJ/m}^3 \, \text{K at standard density}$$

Viscosity of air

$$= 0.018 \times 10^{-3} \, \text{Ns/m}^2$$

Air condition

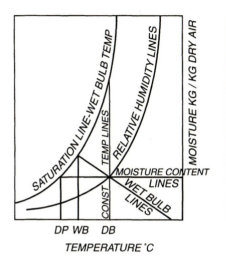

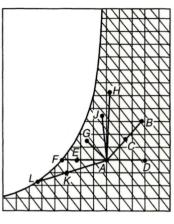

DP WB DB

TEMPERATURE °C

LINES OF CONSTANT ENTHALPY ARE NOT QUITE PARALLEL TO WET BULB LINES
DP   DEW POINT
WB   WET BULB TEMPERATURE
DB   DRY BULB TEMPERATURE

Change of condition of air	Indicated in sketch above	Remarks
Mixing of air volume V_A at condition A	A–C	$\dfrac{\text{Distance AC}}{\text{Distance BC}} = \dfrac{\text{Volume } V_B}{\text{Volume} V_A}$
with air volume V_B at condition B	B–C	
Heating	A–D	
Cooling	A–E-F	Dewpoint at F
Humidification with water injection	A–G	Slope depends on temperature of water but is approximately equal to slope of wet bulb line
Humidification with steam injection	A–H	Temperature increases slightly but for practical purposes can be assumed to be constant
Constant temperature	A–J	
Cooling with dehumidification	A–K	Coil dewpoint L
		Coil contact factor $= \dfrac{DB_A - DB_k}{DB_A - DB_L}$

Relative humidity in per cent

For various room temperatures and various differences between wet and dry bulb temperatures

Dry bulb temp. °C	Difference between dry bulb and wet bulb temperatures (°C)											
	0	1	2	3	4	5	6	7	8	9	10	11
10	100	88	77	66	55	44	34	25	15	6	0	0
11	100	89	78	67	56	46	36	27	18	9	2	0
12	100	89	78	68	58	48	39	29	21	12	3	0
13	100	90	79	69	60	50	41	32	34	15	6	1
14	100	90	80	70	61	52	43	34	26	17	8	2
15	100	90	81	71	62	53	44	36	28	20	11	5
16	100	90	81	71	63	54	46	37	30	22	14	8
17	100	91	82	72	64	56	48	39	32	25	17	11
18	100	91	82	73	65	57	49	41	34	27	20	13
19	100	91	83	74	66	59	51	43	36	29	23	16
20	100	91	83	74	67	59	52	44	38	31	25	18
21	100	92	83	75	68	61	53	46	40	33	27	20
22	100	92	83	75	68	61	54	47	41	34	28	22
23	100	92	84	76	69	62	55	48	42	36	30	24
24	100	92	84	76	70	63	56	49	43	37	31	26
25	100	92	85	77	71	64	57	51	45	39	33	28
26	100	92	85	77	71	64	58	52	46	40	34	29
27	100	93	85	78	72	65	59	53	47	42	36	31
28	100	93	85	78	72	66	59	54	48	43	37	32
29	100	93	86	79	73	67	60	55	49	44	39	34
30	100	93	86	79	73	67	61	56	50	45	40	35
31	100	93	86	80	74	68	62	57	51	46	41	36
32	100	93	86	80	74	68	63	57	52	47	42	37
33	100	93	87	81	75	69	64	58	53	48	43	38
34	100	93	87	81	75	69	64	59	53	49	44	39
35	100	94	87	82	76	70	65	60	54	50	45	40
36	100	94	87	82	76	70	65	60	55	50	46	41
37	100	94	88	82	76	71	66	61	56	51	47	42
38	100	94	88	82	76	71	66	61	56	52	47	43

Relative humidity in per cent

For various room temperatures and various differences between wet
and dry bulb temperatures

| Dry bulb temp. (°F) | Difference between dry bulb and wet bulb temperature (°F) | | | | | | | | | | |
	0	2	4	6	8	10	12	14	16	18	20
50	100	87	74	62	50	39	28	17	7	0	0
52	100	88	75	63	52	41	30	20	10	0	0
54	100	88	76	65	54	43	33	23	14	5	0
56	100	88	77	66	55	45	35	26	17	8	0
58	100	88	77	67	57	47	38	28	20	11	3
60	100	89	78	68	58	49	40	31	22	14	6
62	100	89	79	69	60	50	41	33	25	17	9
64	100	90	79	70	61	52	43	35	27	20	12
66	100	90	80	71	62	53	45	37	29	22	15
68	100	90	81	72	63	55	47	39	31	24	17
70	100	90	81	72	64	56	48	40	33	26	20
72	100	91	82	73	65	57	49	42	35	28	22
74	100	91	82	74	66	58	51	44	37	30	24
76	100	91	83	74	67	59	52	45	38	32	26
78	100	91	83	75	67	60	53	46	40	34	28
80	100	91	83	76	68	61	54	47	41	35	29
82	100	92	84	76	69	62	55	49	43	37	31
84	100	92	84	77	70	63	56	50	44	38	32
86	100	92	85	77	70	63	57	51	45	39	34
88	100	92	85	78	71	64	58	52	46	41	35
90	100	92	85	78	71	65	59	53	47	42	37
92	100	92	85	78	72	65	59	54	48	43	38
94	100	93	86	79	72	66	60	54	49	44	39
96	100	93	86	79	73	67	61	55	50	45	40
98	100	93	86	79	73	67	61	56	51	46	41
100	100	93	86	80	74	68	62	57	52	47	42

Mixture of air and saturated water vapour

Temp. °C	Pressure of sat. vapour kN/m³	Mass of sat. vapour per m³ of mixture g/m³	Mass of sat. vapour per kg of dry air g/kg	Vol. of vapour of dry air m³/kg	Vol. of vapour of mixture m³/kg	Specific entropy of sat. vapour kJ/kg
−15	0.160	1.6	1.0	0.731	0.732	−12.6
−10	0.266	2.3	1.6	0.745	0.746	−6.1
−5	0.399	3.4	2.5	0.759	0.761	+1.09
0	0.612	4.9	3.8	0.773	0.775	9.4
1	0.652	5.2	4.1	0.776	0.778	11.3
2	0.705	5.6	4.4	0.779	0.781	12.9
3	0.758	6.0	4.7	0.782	0.784	14.7
4	0.811	6.4	5.0	0.784	0.787	16.6
5	0.865	6.8	5.4	0.787	0.790	18.5
6	0.931	7.3	5.8	0.791	0.793	20.5
7	0.998	7.7	6.2	0.793	0.796	22.6
8	1.06	8.3	6.7	0.796	0.800	24.7
9	1.14	8.8	7.1	0.799	0.802	26.9
10	1.22	9.4	7.6	0.801	0.805	29.2
11	1.30	10	8.2	0.805	0.808	31.5
12	1.40	11	8.8	0.807	0.812	34.1
13	1.49	11	9.4	0.810	0.814	36.6
14	1.60	12	10.0	0.813	0.818	39.2
15	1.70	13	10.6	0.816	0.821	41.8
16	1.81	14	11.4	0.818	0.824	44.8
17	1.93	14	12.1	0.822	0.828	47.7
18	2.06	15	12.9	0.824	0.831	50.7
19	2.19	16	13.8	0.827	0.833	54.0
20	2.33	17	14.7	0.830	0.837	57.8
21	2.49	18	15.6	0.833	0.840	61.1
22	2.63	19	16.6	0.835	0.844	64.1
23	2.81	20	17.7	0.838	0.847	67.8
24	2.98	22	18.8	0.841	0.850	72.0
25	3.17	23	20.0	0.844	0.854	75.8
26	3.35	24	21.4	0.847	0.858	80.4
27	3.55	26	22.6	0.850	0.861	84.6
28	3.78	27	24.0	0.853	0.865	89.2
29	3.99	29	25.6	0.855	0.869	94.3
30	4.23	30	27.2	0.858	0.873	99.6
35	5.61	39	36.6	0.873	0.892	129
40	7.35	51	48.8	0.887	0.912	166
45	9.56	65	65.0	0.901	0.935	213
50	12.3	82	86.2	0.915	0.959	273
55	15.7	104	114	0.929	0.987	352
60	19.9	130	152	0.943	1.020	456
65	24.9	161	204	0.958	1.057	599

Mixture of air and saturated water vapour

Temp °F	Press. of sat. vapour in Hg	Weight of sat. vapour Grains per ft³	per lb of dry air. Grains per lb	Volume in ft³ of 1 lb of dry air	of 1 lb of dry air & vapour to saturate	Enthalpy of mixture Btu/lb
0	0.0375	0.472	5.47	11.58	11.59	0.852
10	0.00628	0.772	9.16	11.83	11.58	3.831
20	0.01027	1.238	15.01	12.09	12.13	7.137
30	0.1646	1.943	24.11	12.34	12.41	10.933
32	0.1806	2.124	26.47	12.39	12.47	11.83
33	0.1880	2.206	27.57	12.41	12.49	12.18
34	0.1957	2.292	28.70	12.44	12.52	12.60
35	0.2036	2.380	29.88	12.47	12.55	13.02
36	0.2119	2.471	31.09	12.49	12.58	13.44
37	0.2204	2.566	32.35	12.52	12.61	13.87
38	0.2292	2.663	33.66	12.54	12.64	14.31
39	0.2384	2.764	35.01	12.57	12.67	14.76
40	0.2478	2.868	36.41	12.59	12.70	15.21
41	0.2576	2.976	37.87	12.62	12.73	15.67
42	0.2678	3.087	39.38	12.64	12.76	16.14
43	0.2783	3.201	40.93	12.67	12.79	16.62
44	0.2897	3.319	42.55	12.69	12.82	17.10
45	0.3003	3.442	44.21	12.72	12.85	17.59
46	0.3120	3.568	45.94	12.74	12.88	18.09
47	0.3240	3.698	47.73	12.77	12.91	18.60
48	0.3364	3.832	49.58	12.79	12.94	19.12
49	0.3492	3.970	51.49	12.82	12.97	19.65
50	0.3624	4.113	53.47	12.84	13.00	20.19
51	0.3761	4.260	55.52	12.87	13.03	20.74
52	0.3903	4.411	57.64	12.89	13.07	21.30
53	0.4049	4.568	59.83	12.92	13.10	21.87
54	0.4200	4.729	62.09	12.95	13.13	22.45
55	0.4356	4.895	64.43	12.97	13.16	23.04
56	0.4517	5.066	66.85	13.00	13.20	23.64
57	0.4684	5.242	69.35	13.02	13.23	24.25
58	0.4855	5.424	71.93	13.05	13.26	24.88
59	0.5032	5.611	74.60	13.07	13.30	25.52
60	0.5214	5.804	77.30	13.10	13.33	26.18

Mixture of air and saturated water vapour
(*continued*)

Temp °F	Pressure of sat. vapour in Hg	Weight of sat. vapour		Volume in ft³		Enthalpy of mixture Btu/lb
		Grains per ft³	per lb of dry air. Grains per lb	of 1 lb of dry air	of 1 lb of dry air & vapour to saturate	
61	0.5403	6.003	80.2	13.12	13.36	26.84
62	0.5597	6.208	83.2	13.15	13.40	27.52
63	0.5798	6.418	86.2	13.17	13.43	28.22
64	0.6005	6.633	89.3	13.20	13.47	28.93
65	0.6218	6.855	92.6	13.22	13.50	29.65
66	0.6438	7.084	95.9	13.25	13.54	30.39
67	0.6664	7.320	99.4	13.27	13.58	31.15
68	0.6898	7.563	103.0	13.30	13.61	31.92
69	0.7139	7.813	106.6	13.32	13.65	32.71
70	0.7386	8.069	110.5	13.35	13.69	33.51
71	0.7642	8.332	114.4	13.38	13.73	34.33
72	0.7906	8.603	118.4	13.40	13.76	35.17
73	0.8177	8.882	122.6	13.43	13.80	36.03
74	0.8456	9.168	126.9	13.45	13.84	36.91
75	0.8744	9.46	131.4	13.48	13.88	37.81
76	0.9040	9.76	135.9	13.50	13.92	38.73
77	0.9345	10.07	140.7	13.53	13.96	39.67
78	0.9658	10.39	145.6	13.55	14.00	40.64
79	0.9981	10.72	150.6	13.58	14.05	41.63
80	1.0314	11.06	155.8	13.60	14.09	42.64
85	1.212	12.89	184.4	13.73	14.31	48.04
90	1.421	14.96	217.6	13.86	14.55	54.13
95	1.659	17.32	256.3	13.98	14.80	61.01
100	1.931	19.98	301.3	14.11	15.08	68.79
105	2.241	22.99	354	14.24	15.39	77.63
110	2.594	26.38	415	14.36	15.73	87.69
115	2.993	31.8	486	14.49	16.10	99.10
120	3.444	34.44	569	14.62	16.52	112.37
125	3.952	39.19	667	14.75	16.99	127.54
130	4.523	44.49	780	14.88	17.53	145.06
135	5.163	50.38	913	15.00	18.13	165.34
140	5.878	56.91	1072	15.13	18.84	189.22
150	7.566	72.10	1485	15.39	20.60	250.30

Man and air

(a) **Respiration.** An adult at rest breathes 16 respirations per minute, about 0.5 m^3/hr (about 17.5 ft^3hr). When working the rate is 3 to 6 times more.

Average composition of exhaled air
Oxygen   16.5%
Carbon dioxide   4.0%
Nitrogen and argon   79.5%
Quantity of carbon dioxide exhaled in 24 hrs is about 1 kg (2.2 lb).

(b) **Equilibrium of Heat.** Heat is generated within the human body by combustion of food. Heat is lost from the human body by

1 Conduction and convection   about 25%
2 Radiation   about 43%
3 Evaporation of moisture   about 30%
4 Exhaled air   about 2%

Evaporation prevails at high ambient temperatures. Conduction and convection prevail at low ambient temperatures.
Heat is liberated at a rate such that the internal body temperature is maintained at 37°C (98.6 °F).

Proportion of sensible and latent heat dissipated by man at fairly hard work

Dry bulb temp	°C	13	15	18	21	24	27	30	32
	°F	55	60	65	70	75	80	85	90
Sensible heat	%	75	68	60	51	42	31	20	10
Latent heat	%	25	32	40	49	58	69	80	90

(c) **Heat Loss of Human Body.** The total heat loss of an adult (sensible and latent) is approximately 117 W at room temperatures between 18°C and 30°C (about 400 Btu/hr).

Thermal indices are combinations of air temperature, radiant temperature, air movement and humidity to give a measure of a person's feeling of warmth.

(i) *Equivalent temperature* combines the effects of air temperature, radiation and air movement. Numerically it is the temperature of a uniform enclosure in which a sizeable black body maintained at 24°C in still air would lose heat at the same rate as in the environment under consideration. It is measured by a Eupatheoscope.

(ii) *Effective temperature* is an arbitrary index on the basis of subjective assessments of the degree of comfort felt by people in various environments. It takes into account air temperature, air movement and humidity. Numerically it is the temperature of still, saturated air which would produce an identical degree of comfort.

(iii) *Globe temperature* combines the effects of air temperature, radiation and air movement. Numerically it is the reading of a thermometer with its bulb at the centre of a blackened globe 150 mm dia. It is similar to the equivalent temperature but easier to measure.

(iv) *Dry resultant temperature* is similar to globe temperature but the globe used is 100 mm dia. This makes it rather less sensitive to radiation.

(v) *Environmental temperature* combines air temperature and radiation. Numerically it is given by the formula

$$t_{el} = \tfrac{2}{3} t_r + \tfrac{1}{3} t_a$$

where t_{el} = environmental temperature, °C
  t_r = mean radiant temperature of surroundings, °C
  t_a = air temperature, °C

It is not very different from the other scales when air velocity is low and air and radiant temperatures are not widely different, and is easier to use in calculations.

Atmospheric data. Composition of air

Dry air is a mechanical mixtures of gases.

	Dry air per cent		Atmospheric at sea level
	By volume	By weight	By volume
Oxygen	21.00	23.2	20.75
Nitrogen	78.03	75.5	77.08
Carbon dioxide	0.03	0.046	0.03
Hydrogen	0.01	0.007	0.01
Rare gases	0.93	1.247	0.93
Water vapour	–	–	1.20

The composition of air is unchanged to a height of approximately 10 000 metres. The average air temperature diminishes at the rate of about 0.6°C for each 100 m of vertical height.

Altitude-density tables for air

Altitude m	Barometer mm Hg	Altitude m	Barometer mm Hg	Altitude m	Barometer mm Hg
0	749	600	695	1,350	632
75	743	750	681	1,500	620
150	735	900	668	1,800	598
250	726	1,000	658	2,100	577
300	723	1,200	643	2,400	555
450	709				

Altitude ft	Barometer in Hg	Altitude ft	Barometer in Hg	Altitude ft	Barometer in Hg
0	29.92	2,000	27.72	4,500	25.20
250	29.64	2,500	27.20	5,000	24.72
500	29.36	3,000	26.68	6,000	23.79
750	29.08	3,500	26.18	7,000	22.90
1,000	28.80	4,000	25.58	8,000	22.04
1,500	28.31				

Normal Temperature and Pressure (NTP) is 0°C and 101.325 kN/m². Standard Temperature and Pressure (STP) used for determination of fan capacities is 20°C and 101.6 kN/m² or 60°F and 30 in Hg. (These two sets of conditions do no convert directly, but the density of dry air is 1.22 kg/m³ = 0.0764 lb/ft³ at both conditions.)

6 Heat losses

Heat input has to balance heat loss by

1 conduction and convection through walls, windows, etc.
2 infiltration of cold air.

1 Heat loss through walls, windows, doors, ceilings, floors, etc.

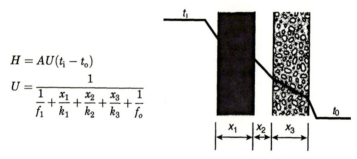

$$H = AU(t_i - t_o)$$

$$U = \cfrac{1}{\dfrac{1}{f_1} + \dfrac{x_1}{k_1} + \dfrac{x_2}{k_2} + \dfrac{x_3}{k_3} + \dfrac{1}{f_o}}$$

where

H = heat transmitted (W)
A = area of exposed surface (m²)
U = overall coefficient of heat transmission (W/m² K)
t_i = inside air temperature (°C)
t_o = outside air temperature (°C)
x = thickness of material (m)
k = thermal conductivity of material (W/m K)
f_i = surface conductance for inside wall (W/m² K)
f_o = surface conductance for outside wall (W/m² K)
$C = \dfrac{k}{x}$ = conductance = heat flow through unit area in unit time (W/m² K)
$R = \dfrac{x}{k} = \dfrac{1}{C}$ = thermal resistivity

2 Heat loss by infiltration

$$H = sdnV(t_1 - t_o)$$

where

H = heat loss (kW)
s = specific heat capacity of air (kJ/kg K)
d = density of air (kg/m^3)
n = number of air changes (1/s)
V = volume of room (m^3)
t_1 = inside air temperature (°C)
t_o = outside air temperature (°C)

Safety additions to heat loss calculations

1   For aspect
North East, 10%.
West, 5%

2   For exposure 5%–10% for surfaces exposed to wind
3   For intermittent heating
Buildings heated during day only. 10–15%
Buildings not in use daily, 25–30%
Buildings with long periods between use (e.g. churches), up to 50%
4   For height

Height of room	m	5	6	7	8	9	10	11	12 and more
Addition	%	2.5	5	7.5	10	12.5	15	17.5	20

Air movement. Air movement makes any conditions of temperature and humidity feel colder; it lowers the effective temperature. An air velocity of 0.12 m/s may be considered as practically still air. A slight air movement is desirable for comfort to remove layers of humid and warm air from the surface of the human body. A higher air velocity is required in air at high temperature and high relative humidity than in air at low temperature and low relative humidity.

Entering air temperature in plenum heating systems must not be too much above or below the room temperature.

For heating
normally   air entering temperature 26–32°C
with good mixing   air entering temperature 38–49°C

For cooling
inlets near occupied zones   5–9°C below room temperature
high velocity jets, diffusion nozzles   17°C below room temperature

Allowance for warming up

(a) rooms heated daily (not at night)

$$H = \frac{0.063(n-1)H_o}{Z} \, \text{W}$$

(b) rooms not heated daily

$$H = \frac{0.1(Z+8)H_t}{Z} \, \text{W}$$

where

H = heat required for warming up (W)
H_o = heat loss through outside surface (W)
H_t = total heat loss (W)
n = interruption of heating (hr)
Z = warming up time (hr)

Air temperatures at various levels

Increase of temperature from 1.5 m to 6 m is at the rate of 7% of temperature at 1.5 m per m. No further increase after 6 m.

$$t' = t + 0.07(h-1.5)t$$

t' = temperature at given level above floor (°C)
t = temperature of 1.5 m above floor (°C)
h = height of given level above floor (m)

Temperature of unheated spaces

$$t = \frac{t_i A_c U_c + t_o A_r U_r}{A_c U_c + A_r U_r}$$

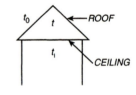

where

t = temperature of unheated space (°C)
t_i = temperature of adjacent room (°C)
t_o = outside temperature (°C)
A_c = area of surface between space and adjacent room – ceiling (m²)
A_r = area of surface between space and outside – roof (m²)
U_c = coefficient of heat transmission between space and adjacent room (W/m² K)
U_r = coefficient of heat transmission between space and outside (W/m² K)

Combined coefficient for ceiling and roof

$$U_E = \frac{U_R U_c}{U_R + \frac{U_c}{r}}$$

where

U_E = combined coefficient of heat transmission from inside to outside, based on ceiling area (W/m² K)
U_R = coefficient of heat transmission of roof (W/m² K)
U_c = coefficient of heat transmission of ceiling (W/m² K)
r = ratio of roof area to ceiling area (dimensionless)

Design winter indoor temperatures (°C)

Heated rooms			
Bars	18	Libraries	20
Bathrooms	22	Living rooms	21
Bedrooms	18	Museums	20
Changing rooms	22	Offices	20
Churches	18	Operating theatres	24
Cloakrooms	16	Prisons	18
Classrooms	20	Recreation rooms	18
Corridors	16	Restaurants	18
Dining rooms	20	Shops	18
Dressing rooms	21	Stores	15
Exhibition halls	18	Swimming baths	27
Factories		Waiting rooms	18
sedentary work	18	Wards	18
light work	16	Warehouses	16
heavy work	13		
Gyms	15	*Unheated rooms*	
Halls,		Attics	0
assembly	18	Attics under	
entrance	16	insulated roof	4
Hotel rooms	21	Cellars	0
Kitchens	16	Foyers with doors	
Laboratories	20	frequently opened	0
Lecture rooms	20	not frequently opened	4
		Internal rooms	2

Design winter outdoor temperatures

For England −4°C to 0°C.

Design infiltration rates

	Air changes per hour		Air changes per hour
Bars	1	Laboratories	1
Bathrooms	2	Lecture rooms	$1\frac{1}{2}$
Bedrooms	1	Libraries	$\frac{1}{2}$
Changing rooms	$\frac{1}{2}$	Living rooms	$1\frac{1}{2}$
Churches	$\frac{1}{2}$	Museums	1
Cloakrooms	1	Offices	1
Classrooms	2	Operating theatres	$\frac{3}{4}$
Corridors	$1\frac{1}{2}$	Prisons	2
Dining rooms	1	Recreation rooms	1
Dressing rooms	1	Restaurants	1
Exhibition halls	$\frac{1}{2}$	Shops	1
Factories	$1-1\frac{1}{2}$	Stores	$\frac{1}{2}$
Gyms	1	Swimming baths	$\frac{1}{2}$
Halls, assembly	$\frac{1}{2}$	Waiting rooms	1
entrance	2	Wards	2
Hotel rooms	1	Warehouses	$\frac{1}{2}$

Typical Air Infiltration Rates
10–27 m^3/hr per m^2 of facade at 50 N/m^2 pressure difference between inside and outside

Heat loss calculations for high buildings

Floor	Addition to infiltration rate	Designation of U-valve
Ground, 1st	nil	Normal
2nd to 4th	25%	Normal
5th to 11th	50%	Normal
Above 11th	100%	Severe

Infiltration heat loss

Heat loss for 1 air change per hour $= 0.34$ W/m^3 K (0.018 Btu/hr ft^3 °F).

Heat loss calculation

Contract temperatures and their equivalents

Inside temperatures obtained with a certain system with outside temperatures other than for which the system is designed. (Empirical formula by J. Roger Preston.)

$$t_4 = \left(t_1^{12} - t_2^{12} + t_3^{12}\right)^{1/12}$$

where

t_1 = Contract inside temperature (K)
t_2 = Contract outside temperature (K)
t_3 = Existing outside temperature (K)
t_4 = Estimated inside temperature (K)

(Formula remains unchanged if all temperatures are in °F absolute.)

Table for 30°F contract outside and 60°F contract inside

Existing Outside temp. °F	20	22	24	26	28	**30**	32	34	36	38	40
Inside temp. °F	55	56	57	58	59	**60**	61	62	63	64	65

Table for 0°C contract outside and 20°C contract inside

Existing Outside temp. °C	−5	−4	−3	−2	−1	**0**	+1	2	3	4	5
Inside temp. °C	17.8	18.3	18.8	19.0	19.6	**20**	20.5	21.0	21.4	21.7	22.5

Thermal conductivities

Material	Conductivity k		Resistivity $1/k$	
	$\dfrac{Btu\ in}{ft^2\ hr\ °F}$	$W/m\ K$	$\dfrac{ft^2\ hr\ °F}{Btu\ in}$	$m\ K/W$
Air	0.18	0.026	5.56	38.6
Aluminium	1050	150		
Asbetolux	0.8	0.12	1.25	8.67
Asbestos:				
flues and pipes	1.9	0.27	0.53	3.68
insulating board	1.0	0.14	1.0	6.93
lightweight slab	0.37	0.053	2.70	18.7
Asphalt:				
light	4.0	0.58	0.25	1.73
heavy	8.5	1.23	0.12	0.83
Brass	550	150		
Bricks:				
common	9.9	1.43	0.10	0.69
engineering	5.5	0.79	0.18	1.25
Brine	3.3	0.48	0.30	2.10
Building				
board	0.55	0.079	1.82	12.62
paper	0.45	0.065	2.22	15.39
Caposite	0.36	0.052	2.78	19.28
Cardboard	1.0 to 2.0	0.144 to 0.288	1.0 to 0.5	6.9 to 3.5
Celotex	0.33	0.048	3.0	21.0
Concrete:				
1:2:4	10.0	1.4	0.10	0.69
lightweight	2.8	0.40	0.36	2.5
Copper	2100	300		
Cork	0.30	0.043	3.33	23.1
Densotape	1.7	0.25	0.58	4.0
Diatomaceous earth	0.60	0.087	1.66	11.5
Econite	0.68	0.098	1.47	10.19
Felt	0.27	0.039	3.70	25.7
Fibreglass	0.25	0.036	4.0	27.7
Firebrick	9.0	1.30	0.11	0.76
Fosalsil	1.0	0.14	0.10	0.69
Glass	7.3	1.05	0.14	0.97
Glasswool	0.28	0.04	3.6	24.8
Gold	2150	310		

Thermal conductivities (*continued*)

Material	Conductivity k Btu in / ft^2 hr °F	W/m K	Resistivity $1/k$ ft^2 hr °F / Btu in	m K/W
Granwood floor blocks	2.20	0.32	0.45	3.1
Gyproc plasterboard	1.1	0.16	0.91	6.3
Gypsum plasterboard	1.1	0.16	0.91	6.3
Hardboard	0.65	0.094	1.54	10.68
Holoplast: 25 mm panel	0.95	0.14	1.05	7.3
Ice	16.0	2.31	0.0625	0.43
Insulating board	0.41	0.059	2.45	16.99
Iron:				
cast	450	65	0.0022	0.154
wrought	400	58	0.0025	0.0172
Jute	0.25	0.036	4.0	27.7
Kapok	0.25	0.036	4.0	27.7
Lead	240	35	0.0042	0.029
Linoleum:				
cork	0.5	0.072	2.0	13.9
p.v.c.	1.5	0.22	0.67	4.65
rubber	2.1	0.30	0.48	3.33
Marinite	0.74	0.11	1.35	9.36
Mercury	48	7	0.021	0.143
Mica sheet	4.5	0.65	0.22	1.53
Mineral wool	0.39	0.056	3.33	23.1
Nickel	400	58	0.0025	0.0172
On ozote	0.20	0.029	5.0	34.7
Paper	0.90	0.13	0.11	7.69
Perspex	1.45	0.21	0.69	4.8
Plaster	3.3	0.48	0.30	2.1
Platinum	480	69	0.0021	0.0145
Polystyrene: cellular	0.23	0.033	4.3	29.8
Polyurethane: cellular	0.29	0.042	3.45	23.9
Polyzote	0.22	0.032	4.55	31.5
Porcelain	7.2	1.04	0.14	0.96

Thermal conductivities (*continued*)

Material	Conductivity k		Resistivity $1/k$	
	$\dfrac{Btu\ in}{ft^2\ hr\ °F}$	W/m K	$\dfrac{ft^2\ hr\ °F}{Btu\ in}$	m K/W
Refractory brick alumina	2.2	0.32	0.45	3.1
diatomaceous	0.9	0.13	1.11	7.70
silica	10.0	1.44	0.10	0.69
vermiculite insulating	1.35	0.19	0.74	5.13
Refractory concrete:				
diatomaceous	1.8	0.26	0.56	3.9
aluminous cement	3.2	0.46	0.31	2.15
Rubber:				
natural	1.1	0.16	0.91	6.3
silicone	1.6	0.23	0.63	4.4
Sand	2.9	0.42	0.35	2.4
Scale, boiler	16.0	2.3	0.0625	0.43
Silver	2900	420		
Sisalkraft building paper	0.46	0.066	2.17	15.0
Slate	14.0	2.0	0.071	0.5
Snow	1.5	0.22	0.67	4.65
Steel, soft	320	46		
Steel wool	0.75	0.108	1.33	9.22
Stillite	0.25	0.036	4.0	27.7
Stone:				
granite	20.3	2.9	0.05	0.35
limestone	10.6	1.5	0.09	0.62
marble	17.4	2.5	0.06	0.42
sandstone	13.0	1.9	0.08	0.55
Sundeala:				
insulating board	0.36	0.052	2.78	19.3
medium hardboard	0.51	0.074	2.0	13.9
Tentest	0.35	0.05	2.86	19.8
Thermalite	1.4	0.20	0.71	4.9
Tiles:				
asphalt and asbestos	3.8	0.55	0.26	1.8
burnt clay	5.8	0.84	0.17	1.2
concrete	8.0	1.2	0.13	0.90
cork	0.58	0.084	1.72	11.9
plaster	2.6	0.37	0.38	2.63

Thermal conductivities (*continued*)

Material	Conductivity k		Resistivity $1/k$	
	$\dfrac{Btu\ in}{ft^2\ hr\ °F}$	$W/m\ K$	$\dfrac{ft^2\ hr\ °F}{Btu\ in}$	$m\ K/W$
Timber:				
balsa	0.33	0.048	3.0	20.8
beech	1.16	0.17	0.86	5.97
cypress	0.67	0.097	1.49	10.3
deal	0.87	0.13	1.15	7.97
fir	0.76	0.11	1.3	9.1
oak	1.11	0.16	0.90	6.24
plywood	0.96	0.14	1.04	7.21
teak	0.96	0.14	1.04	7.21
Treetex	0.39	0.056	2.56	17.8
Water	4.15	0.60	0.24	1.7
Weyboard	0.63	0.091	1.60	11.1
Weyroc	1.0	0.14	1.0	6.9
Woodwool	0.28	0.040	3.58	24.8
Wool	0.30	0.043	3.33	23.1
Zinc	440	64		
Sawdust	0.49	0.071	2.04	14.1
Cotton waste	0.41	0.059	2.4	16.9

Thermal transmittance coefficients for building elements

Orientation		Exposure											
S		Sheltered		Normal		Severe		—		—		—	
W SW SE		—		Sheltered		Normal		Severe		—		—	
NW		—		—		Sheltered		Normal		Severe		—	
N NE E		—		—		—		Sheltered		Normal		Severe	
		$Btu/ft^2\ hr\,°F$	$W/m^2\ K$	$Btu/ft^2\ hr\,°F$	$W/m^2\ K$	$Btu/ft^2\ hr\,°F$	$W/m^2\ K$	$Btu/ft^2\ hr\,°F$	$W/m^2\ K$	$Btu/ft^2\ hr\,°F$	$W/m^2\ K$	$Btu/ft^2\ hr\,°F$	$W/m^2\ K$

Walls

Solid brick — Unplastered

Thickness													
100 mm		0.50	2.9	0.55	3.1	0.59	3.4	0.64	3.6	0.69	3.9	0.75	4.3
225 mm		0.39	2.2	0.42	2.4	0.44	2.5	0.47	2.7	0.50	2.9	0.53	3.0
340 mm		0.32	1.8	0.34	1.9	0.35	2.0	0.37	2.1	0.39	2.2	0.41	2.3

Solid brick — Plastered

Thickness													
100 mm		0.46	2.6	0.49	2.8	0.53	3.0	0.57	3.2	0.61	3.5	0.65	3.7
225 mm		0.36	2.1	0.38	2.2	0.41	2.3	0.43	2.4	0.45	2.6	0.48	2.7
340 mm		0.30	1.7	0.32	1.8	0.33	1.9	0.35	2.0	0.36	2.1	0.38	2.2
455 mm		0.26	1.5	0.27	1.5	0.28	1.6	0.29	1.6	0.30	1.7	0.31	1.8
560 mm		0.23	1.3	0.23	1.3	0.24	1.4	0.25	1.4	0.26	1.5	0.26	1.5

Cavity brick — Unventilated

Thickness													
270 mm		0.27	1.5	0.28	1.6	0.29	1.6	0.30	1.7	0.31	1.8	0.32	1.8
390 mm		0.23	1.3	0.24	1.4	0.25	1.4	0.26	1.5	0.27	1.5	0.27	1.5
500 mm		0.21	1.2	0.21	1.2	0.22	1.2	0.22	1.2	0.23	1.3	0.24	1.4

Cavity brick — Ventilated

Thickness													
270 mm		0.30	1.7	0.31	1.8	0.33	1.9	0.34	1.9	0.36	2.0	0.37	2.1
390 mm		0.26	1.5	0.27	1.5	0.28	1.6	0.29	1.6	0.30	1.7	0.31	1.8
500 mm		0.22	1.2	0.23	1.3	0.24	1.4	0.25	1.4	0.25	1.4	0.26	1.5

Cavity brick — Plastered

Thickness													
200 mm		0.31	1.8	0.32	1.8	0.34	1.9	0.36	2.0	0.37	2.1	0.39	2.2
270 mm		0.23	1.3	0.23	1.3	0.23	1.3	0.24	1.4	0.25	1.4	0.26	1.5
390 mm		0.18	1.0	0.19	1.1	0.19	1.1	0.20	1.1	0.20	1.1	0.21	1.2

Concrete

Thickness													
100 mm		0.55	3.1	0.60	3.4	0.66	3.8	0.71	4.0	0.78	4.4	0.85	4.8
150 mm		0.49	2.8	0.53	3.0	0.58	3.3	0.63	3.6	0.68	3.9	0.73	4.1
200 mm		0.45	2.5	0.48	2.7	0.52	3.0	0.56	3.2	0.60	3.4	0.64	3.6
250 mm		0.41	2.3	0.44	2.5	0.47	2.7	0.50	2.8	0.53	3.0	0.57	3.2

Wood	25 mm	0.41	2.3	0.44	2.5	0.47	2.7	0.50	2.8	0.53	3.0	0.56	3.2
Tongued and grooved	38 mm	0.34	1.9	0.36	2.0	0.38	2.2	0.40	2.3	0.42	2.4	0.44	2.5
Walls													
Asbestos sheeting	6 mm	0.64	3.1	0.72	4.1	0.80	4.6	0.89	5.1	1.00	5.7	1.12	6.4
Corrugated iron	1.6 mm	0.79	4.5	0.91	5.2	1.04	5.9	1.20	6.8	1.40	8.0	1.67	9.5
Stone	300 mm	0.41	2.3	0.44	2.5	0.47	2.7	0.50	2.8	0.53	3.0	0.56	3.2
	450 mm	0.34	1.9	0.36	2.0	0.38	2.2	0.40	2.3	0.42	2.4	0.44	2.5
	600 mm	0.29	1.6	0.31	1.8	0.32	1.8	0.33	1.9	0.35	2.0	0.36	2.0
Cavity, inner leaf 100 mm thermalite, outer leaf 100 mm brick, 50 mm cavity		0.18	1.0	0.18	1.0	0.19	1.1	0.19	1.1	0.19	1.1	0.21	1.2
Cavity, inner leaf thermalite 100 mm, outer leaf brick 100 mm, air gap 50 mm, lined internally plasterboard		0.16	0.92	0.17	0.95	0.17	0.97	0.17	0.99	0.18	1.0	0.18	1.0
Insulated, inner leaf 100 mm brick, 50 mm Polyurethane foam, outer leaf 100 mm brick internally plastered		0.31	1.8	0.32	1.8	0.34	1.9	0.36	2.0	0.37	2.1	0.39	2.2

Thermal transmittance coefficients for building elements (continued)

Orientation	Exposure											
S	Sheltered		Normal		Severe		—		—		—	
W SW SE	—		Sheltered		Normal		Severe		—		—	
NW	—		—		Sheltered		Normal		Severe		—	
N NE E	—		—		—		Sheltered		Normal		Severe	
	$Btu/ft^2\,hr\,°F$	$W/m^2\,K$	$Btu/ft^2\,hr\,°F$	$W/m^2\,K$	$Btu/ft^2\,hr\,°F$	$W/m^2\,K$	$Btu/ft^2\,hr\,°F$	$W/m^2\,K$	$Btu/ft^2\,hr\,°F$	$W/m^2\,K$	$Btu/ft^2\,hr\,°F$	$W/m^2\,K$
Insulated, outer leaf 110 mm brick, 50 mm cavity, inner leaf 110 mm lightweight concrete, 50 mm fibreglass insulation, lined with plasterboard	0.081	0.45	0.082	0.45	0.083	0.45	0.084	0.46	0.084	0.46	0.085	0.47
as above, 75 mm fibreglass	0.062	0.34	0.062	0.35	0.063	0.35	0.063	0.35	0.064	0.35	0.064	0.36
as above, 100 mm fibreglass	0.050	0.28	0.050	0.28	0.051	0.28	0.051	0.28	0.051	0.28	0.051	0.29
Insulated, outer leaf 110 mm brick, inner leaf 110 mm lightweight concrete lined with plaster or plasterboard, insulation between leaves 50 mm fibreglass	0.088	0.49	0.089	0.49	0.090	0.49	0.091	0.50	0.092	0.51	0.093	0.52
as above, 75 mm fibreglass	0.065	0.36	0.066	0.37	0.066	0.37	0.067	0.37	0.068	0.38	0.068	0.38
as above, 50 mm polystyrene	0.083	0.46	0.084	0.47	0.085	0.47	0.086	0.48	0.087	0.48	0.087	0.48
as above, 75 mm polystyrene	0.062	0.34	0.062	0.35	0.063	0.35	0.063	0.35	0.064	0.35	0.064	0.36

Outer leaf 110 mm brick inner leaf 110 mm thermalite, lined with plaster or plasterboard, insulation between leaves												
50 mm fibreglass	0.080	0.44	0.081	0.45	0.081	0.45	0.082	0.46	0.083	0.46	0.084	0.46
as above, 75 mm fibreglass	0.061	0.34	0.061	0.34	0.062	0.34	0.062	0.35	0.063	0.35	0.063	0.35
as above, 50 mm polystyrene	0.076	0.42	0.077	0.43	0.078	0.43	0.078	0.44	0.079	0.44	0.080	0.44
as above, 75 mm polystyrene	0.058	0.32	0.058	0.32	0.059	0.33	0.059	0.33	0.060	0.33	0.060	0.33
Outer leaf 150 mm concrete inner leaf 225 mm thermalite, plastered, insulation between leaves												
50 mm fibreglass	0.062	0.35	0.062	0.35	0.063	0.36	0.063	0.36	0.064	0.36	0.064	0.37
as above 75 mm fibreglass	0.050	0.28	0.050	0.28	0.050	0.29	0.051	0.29	0.051	0.29	0.051	0.29
Outer leaf 150 mm concrete inner leaf 150 mm lightweight concrete, plastered, insulation between leaves												
50 mm fibreglass	0.083	0.47	0.084	0.48	0.085	0.48	0.086	0.49	0.087	0.49	0.088	0.50
as above 75 mm fibreglass	0.063	0.36	0.063	0.36	0.064	0.36	0.065	0.37	0.065	0.37	0.065	0.37

Thermal transmittance coefficients for building elements (continued)

Exposure / **Orientation**

Orientation–exposure key (as printed):

Orientation	Exposure
S	Sheltered
W SW SE	Normal
NW	Severe
N NE E	Severe

	Sheltered Btu/ft² hr°F	Sheltered W/m² K	Normal Btu/ft² hr°F	Normal W/m² K	Severe Btu/ft² hr°F	Severe W/m² K	Sheltered Btu/ft² hr°F	Sheltered W/m² K	Normal Btu/ft² hr°F	Normal W/m² K	Severe Btu/ft² hr°F	Severe W/m² K
Walls, concrete, 250 mm concrete, lined internally with 50 mm fibreglass and plasterboard	0.096	0.53	0.098	0.54	0.099	0.55	0.10	0.56	0.10	0.56	0.10	0.57
as above, 75 mm fibreglass	0.070	0.39	0.071	0.39	0.072	0.40	0.072	0.40	0.073	0.41	0.073	0.41
as above, 50 mm polystyrene	0.091	0.51	0.092	0.51	0.094	0.52	0.095	0.53	0.096	0.53	0.096	0.54
as above, 75 mm polystyrene	0.066	0.37	0.067	0.37	0.067	0.37	0.068	0.38	0.069	0.38	0.069	0.38
Windows												
Single glazed	0.70	4.0	0.79	4.5	0.88	5.0	1.00	5.7	1.14	6.5	1.30	7.4
Double glazed												
20 mm air gap	0.41	2.3	0.44	2.5	0.47	2.7	0.50	2.8	0.53	3.0	0.56	3.2
12 mm air gap	0.44	2.4	0.47	2.6	0.51	2.9	0.54	2.9	0.58	3.1	0.62	3.3
6 mm air gap	0.47	2.7	0.51	2.9	0.54	3.1	0.58	3.3	0.63	3.6	0.67	3.8
3 mm air gap	0.52	2.9	0.57	3.2	0.61	3.5	0.68	3.9	0.73	4.1	0.79	4.5
Triple glazed												
20 mm air gap	0.29	1.6	0.31	1.8	0.32	1.8	0.33	1.9	0.35	2.0	0.36	2.0
12 mm air gap	0.32	1.7	0.34	1.8	0.35	2.0	0.37	2.0	0.39	2.1	0.41	2.2
6 mm air gap	0.35	2.0	0.37	2.1	0.39	2.2	0.41	2.3	0.43	2.4	0.46	2.6
3 mm air gap	0.41	2.3	0.44	2.5	0.47	2.7	0.50	2.7	0.54	2.8	0.57	3.2

Internal floors and ceilings	$\dfrac{Btu}{ft^2 hr\,°F}$	W/m^2K
25 mm screed on 150 mm concrete	0.48	2.7
50 mm screed on 150 mm concrete	0.46	2.6
25 mm floorboards on joists, plastered ceiling	0.080	0.45
25 mm floorboards on joists, plasterboard ceiling	0.079	0.45

Internal partitions	$\dfrac{Btu}{ft^2 hr\,°F}$	W/m^2K
110 mm brick	0.55	3.1
225 mm brick	0.44	2.5
340 mm brick	0.37	2.1
110 mm brick, plastered both sides	0.53	3.0
225 mm brick, plastered both sides	0.43	2.4
340 mm brick, plastered both sides	0.36	2.0
100 mm thermalite, plastered both sides	0.24	1.3
225 mm thermalite, plastered both sides	0.13	0.73
100 mm lightweight concrete, plastered both sides	0.35	2.0
150 mm lightweight concrete, plastered both sides	0.28	1.6
250 mm lightweight concrete, plastered both sides	0.20	1.1
Plasterboard, air gap, plasterboard	0.36	2.1

	Exposure					
	Sheltered		Normal		Severe	
	$\dfrac{Btu}{ft^2\,hr\,°F}$	$\dfrac{W/m^2}{K}$	$\dfrac{Btu}{ft^2\,hr\,°F}$	$\dfrac{W/m^2}{K}$	$\dfrac{Btu}{ft^2\,hr\,°F}$	$\dfrac{W/m^2}{K}$
Flat roofs						
Asphalt on 150 mm concrete	0.58	3.3	0.64	3.6	0.70	4.0
Asphalt on 150 mm concrete with plaster underneath	0.51	2.9	0.55	3.1	0.61	3.5
Asphalt on 150 mm hollow tiles	0.45	2.5	0.48	2.7	0.52	3.0
Asphalt on 150 mm hollow tiles with lightweight screed and plaster underneath	0.30	1.7	0.32	1.8	0.33	1.9
Asphalt with screed on 50 mm woodwool slabs on timber joists and plaster ceiling	0.16	0.9	0.18	1.0	0.21	1.2
Asphalt with screed on 50 mm woodwool slabs on steel framing	0.24	1.4	0.26	1.5	0.28	1.6
Asphalt on 50 mm screed on 50 mm fibreglass on steel sheet over 50 mm air gap with plasterboard underneath	0.092	0.51	0.094	0.52	0.095	0.53
as above, 75 mm fibreglass	0.068	0.38	0.069	0.38	0.069	0.39
as above, 100mm fibreglass	0.054	0.30	0.054	0.30	0.055	0.30
Asphalt on felt over 50 mm fibreglass on 150 mm concrete	0.10	0.57	0.10	0.58	0.11	0.58
as above, 75 mm fibreglass	0.073	0.41	0.074	0.41	0.075	0.42
as above, 100 mm fibreglass	0.057	0.32	0.058	0.32	0.058	0.32
Stone chippings on steel deck over 50 mm air gap above 50 mm fibreglass on plasterboard	0.098	0.54	0.099	0.55	0.10	0.56
as above, 75 mm fibreglass	0.076	0.39	0.071	0.40	0.072	0.42
as above, 100 mm fibreglass	0.056	0.31	0.056	0.31	0.057	0.31
Pitched roofs						
Corrugated aluminium sheeting	0.90	5.1	1.15	6.6	1.45	8.3
Corrugated steel sheeting	0.90	5.1	1.15	6.6	1.45	8.3
Tiles on battens and roofing felt with rafters and plasterboard ceiling	0.32	1.8	0.35	2.0	0.39	2.2
Tiles on battens and roofing felt with rafters and plasterboard ceiling with boarding on rafters	0.26	1.5	0.30	1.7	0.33	1.9
Tiles on battens and rafters with plasterboard ceiling	0.44	2.5	0.49	2.8	0.53	3.0
Tiles on battens and rafters with plasterboard ceiling and boarding on rafters	0.32	1.8	0.35	2.0	0.39	2.2

		Exposure					
		Sheltered		*Normal*		*Severe*	
		$\dfrac{Btu}{ft^2\,hr\,°F}$	$\dfrac{W/m^2}{K}$	$\dfrac{Btu}{ft^2\,hr\,°F}$	$\dfrac{W/m^2}{K}$	$\dfrac{Btu}{ft^2\,hr\,°F}$	$\dfrac{W/m^2}{K}$
Tiles on battens and roofing felt with rafters and no ceiling below		0.69	3.9	0.76	4.3	0.83	4.7
Tiles on battens and boarding with rafters and no ceiling below		0.53	3.0	0.58	3.3	0.63	3.6
Tiles on battens only with rafters and no ceiling below		1.00	5.7	1.11	6.3	1.23	7.0
Tiles on battens and roofing felt, plasterboard ceiling with fibreglass insulation insulation thickness	50 mm	0.090	0.51	0.090	0.51	0.092	0.52
	75 mm	0.066	0.38	0.067	0.38	0.068	0.38
	100 mm	0.053	0.30	0.053	0.30	0.053	0.30
	150 mm	0.037	0.21	0.037	0.21	0.038	0.21
Tiles on roofing felt and battens with boarding, plasterboard ceiling with fibreglass insulation and boarding insulation thickness	50 mm	0.085	0.48	0.086	0.49	0.087	0.49
	75 mm	0.064	0.36	0.064	0.36	0.065	0.37
	100 mm	0.051	0.29	0.051	0.29	0.052	0.29
	150 mm	0.036	0.21	0.036	0.21	0.037	0.21
Roof glazing							
Skylight		1.00	5.7	1.20	6.8	1.40	8.0
Laylight with lantern over		0.57	3.2	0.60	3.4	0.63	3.6
Filon transluscent GRP							
single skin		0.89	5.0	1.0	5.7	1.20	6.7
double skin		0.47	2.6	0.57	2.8	0.54	3.0

Floors		Solid floor with four exposed edges		Solid floor with two exposed edges		Suspended floor	
Width m	Length m	$\dfrac{Btu}{ft^2\,hr\,°F}$	$\dfrac{W/m^2}{K}$	$\dfrac{Btu}{ft^2hr\,°F}$	$\dfrac{W/m^2}{K}$	$\dfrac{Btu}{ft^2hr\,°F}$	$\dfrac{W/m^2}{K}$
60	over 100	0.016	0.09	0.009	0.05	0.019	0.11
60	100	0.021	0.12	0.012	0.07	0.025	0.14
60	60	0.026	0.15	0.014	0.08	0.028	0.16
40	over 100	0.021	0.12	0.012	0.07	0.026	0.15
40	100	0.026	0.15	0.016	0.09	0.032	0.18
40	60	0.030	0.17	0.018	0.10	0.035	0.20
40	40	0.037	0.21	0.021	0.12	0.039	0.22
20	over 100	0.039	0.22	0.021	0.12	0.046	0.26
20	100	0.042	0.24	0.025	0.14	0.049	0.28
20	60	0.046	0.26	0.026	0.15	0.053	0.30
20	40	0.049	0.28	0.028	0.16	0.055	0.31
20	20	0.063	0.36	0.037	0.21	0.065	0.37
10	100	0.062	0.35	0.039	0.22	0.077	0.44
10	60	0.072	0.41	0.042	0.24	0.081	0.46
10	40	0.076	0.43	0.044	0.25	0.083	0.47
10	20	0.085	0.48	0.049	0.28	0.090	0.51
10	10	0.11	0.62	0.063	0.36	0.10	0.59
6	40	0.10	0.59	0.062	0.35	0.11	0.63
6	20	0.11	0.64	0.067	0.38	0.11	0.65
6	10	0.13	0.74	0.077	0.44	0.13	0.71
6	6	0.16	0.91	0.095	0.54	0.14	0.79
4	40	0.12	0.66	0.069	0.39	0.13	0.71
4	20	0.14	0.82	0.086	0.49	0.14	0.79
4	10	0.16	0.90	0.095	0.54	0.15	0.83
4	6	0.18	1.03	0.11	0.62	0.16	0.89
4	4	0.21	1.22	0.13	0.73	0.17	0.96
2	20	0.18	1.03	0.13	0.75	0.17	0.96
2	10	0.23	1.31	0.14	0.82	0.19	1.08
2	6	0.25	1.40	0.15	0.87	0.20	1.11
2	4	0.27	1.52	0.17	0.95	0.20	1.15
2	2	0.35	1.96	0.21	1.22	0.22	1.27

External resistance R_{S2}

Orientation	Sheltered $\dfrac{ft^2\,hr\,°F}{Btu}$	Sheltered $\dfrac{m^2\,K}{W}$	Normal $\dfrac{ft^2\,hr\,°F}{Btu}$	Normal $\dfrac{m^2\,K}{W}$	Severe $\dfrac{ft^2\,hr\,°F}{Btu}$	Severe $\dfrac{m^2\,K}{W}$
S	0.73	0.128	0.57	0.100	0.43	0.076
W, SW, SE	0.57	0.100	0.43	0.076	0.30	0.053
NW	0.43	0.076	0.30	0.053	0.18	0.032
N, NE, E	0.43	0.076	0.30	0.053	0.07	0.012
Horizontal (roof)	0.40	0.070	0.25	0.044	0.10	0.018

Exposure spans the Sheltered, Normal and Severe column groups.

Internal resistance R_{S1}

	$\dfrac{ft^2\,hr\,°F}{Btu}$	$\dfrac{m^2\,K}{W}$
Walls	0.70	0.123
Floors	0.85	0.150
Ceilings and roofs	0.60	0.106

Note: The data for surface resistances are applicable to plain surfaces but not to bright metallic surfaces.

The resistance of a corrugated surface is less than that of a plain one, generally by about 20%.

Thermal resistivity of air spaces

Material bounding space — resistivity (x/k) in $m^2\,K/W$ for thickness of air space in mm

Material bounding space	15	20	25	35	50	65	75	90	100	115
Glass	0.141	0.145	0.148	0.155	0.165	0.172	0.176	0.183	0.186	0.188
Brick	0.150	0.153	0.158	0.165	0.175	0.185	0.190	0.197	0.200	0.203

Material bounding space — resistivity (x/k) in $ft^2\,hr\,°F/Btu$ for thickness of air space in in

Material bounding space	$\frac{1}{2}$	$\frac{3}{4}$	1	$1\frac{1}{2}$	2	$2\frac{1}{2}$	3	$3\frac{1}{2}$	4	$4\frac{1}{2}$	5
Glass	0.79	0.82	0.85	0.89	0.94	0.97	1.00	1.04	1.06	1.07	1.08
Brick	0.84	0.87	0.90	0.95	1.04	1.04	1.08	1.11	1.14	1.16	1.17

Condensation on glass windows

The chart gives the maximum permissible heat transfer coefficient of the glass necessary to prevent condensation at various indoor and outdoor temperatures and humidity.

Example

Inside temp.	15°C
Inside rel. humidity	30%
Outside temps.	−5°C

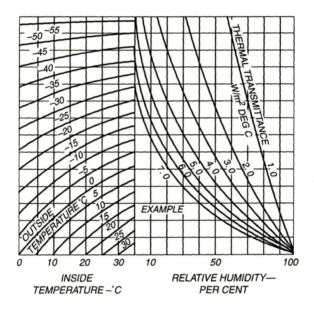

From chart, maximum permissible thermal transmittance coefficient is 7.0 W/m² K.

Fuel consumption

1 Direct method

$$F = \frac{Hn(t_1 - t_a)100}{EC(t_i - t_o)}$$

where

F = fuel consumption during time n (kg)
H = heat loss for temperature difference $(t_i - t_o)$ (kW)
n = time over which fuel consumption is required (s)
E = efficiency of utilisation of fuel (%)
C = calorific value of fuel (kJ/kg)
t_i = inside temperature (°C)
t_a = average outside temperature during period considered (°C)
t_o = outside design temperature (°C)

$$E = E_1\ E_2\ E_3\ E_4$$

where

E_1 = boiler efficiency   (60–75%)
E_2 = efficiency of pipework (loss of heat from pipes)   (80–90%)
E_3 = efficiency of heaters   (90–100%)
E_4 = efficiency of control (loss due to over heating)   (80–95%)
E = efficiency of utilisation of fuel   (35–65%)

2 Degree day method

Degree days give the extent and length of time that the outdoor temperature is below 15.5°C.

number of degree days in a stated period
= number of days × (15.5°C−average outdoor temperature°C)

$$F = \frac{hD100}{EC}$$

$$h = \frac{24 \times 3600 \times H}{(15.5 - t_o)}$$

where

F = fuel consumption over period considered (kg)
h = heat loss per degree day (kJ/degree day)
E = efficiency of utilisation of fuel, as above (%)
C = calorific value of fuel (kJ/kg)
H = heat loss for design conditions (kW)
t_o = outside design temperature (°C)
D = actual number of degree days in period considered (number)

Degree days for United Kingdom
Base temperature 15.5°C

| Region | *Month* | | | | | | | | | | | | |
	Jan.	Feb.	Mar.	Apr.	May	Jun.	Jul.	Aug.	Sep.	Oct.	Nov.	Dec.	Total
Thames Valley	346	304	282	197	113	47	24	27	56	132	256	333	2118
South Eastern	370	329	310	224	145	74	44	48	84	163	280	356	2427
Southern	339	307	294	214	141	68	41	42	76	145	258	328	2253
South Western	293	272	267	197	131	58	32	30	55	114	215	276	1940
Severn Valley	344	311	292	209	129	56	31	34	69	143	259	328	2205
Midland	371	335	318	233	152	76	46	51	92	172	290	358	2494
West Pennines	359	323	304	222	139	66	41	43	79	155	280	346	2357
North Western	366	333	319	239	163	82	55	56	94	169	296	357	2531
Borders	376	343	332	259	193	176	73	72	108	184	300	361	2718
North Eastern	374	334	317	234	154	76	48	50	87	171	295	360	2500
East Pennines	362	323	304	217	139	66	40	42	77	157	281	350	2358
East Anglia	378	334	315	232	143	71	46	43	74	154	283	360	2433
West Scotland	368	335	316	235	163	83	61	63	106	179	303	354	2565
East Scotland	379	343	326	252	189	100	69	71	106	185	308	368	2696
North East Scotland	396	359	345	270	206	111	85	86	124	199	322	381	2884
Wales	323	301	292	228	156	80	49	43	72	136	239	301	2220
Northern Ireland	359	325	311	237	167	86	61	63	100	170	288	343	2510

7 Cooling loads

Cooling load for air conditioning consists of:

conduction and convection through walls, windows, etc.
absorption of solar radiation on walls, windows, etc.
heat emission of occupants.
infiltration of warm outdoor air.
heat emission of lights and other electrical or mechanical appliances.

1 Heat gain through walls, windows, doors, etc.

$$H = AU(t_o - t_i)$$

where

H = Heat gained (W)
A = area of exposed surface (m^2)
U = coefficient of heat transmission (W/m^2 K)
t_o = outside air temperature (°C)
T_i = indoor air temperature (°C)

Coefficients of heat transmission are the same as for heat losses in winter.

2 Solar radiation

$$H = AF\alpha J$$

where

H = heat gained (W)
A = area of exposed surface (m^2)
F = radiation factor
  = proportion of absorbed radiation which is transmitted to interior
α = absorption coefficients
  = proportion of incident radiation which is absorbed
J = intensity of solar radiation striking the surface (W/m^2)

3 Heat emission of occupants

Heat and moisture given off by human body; tabulated data.

4 Heat gain by infiltration

$$H = nVd(h_o - h_1)$$

where

H = heat gain (kW)
n = number of air changes (s^{-1})
V = volume of room (m^3)
d = density of air (kg/m^3)
h_o = enthalpy of outdoor air with water vapour (kJ/kg)
h_1 = enthalpy of indoor air with water vapour (kJ/kg)

5 Heat emission of appliances

All power consumed is assumed to be dissipated as heat.

heat emission in kW = appliance input rating in kW

Lighting	10–14 W/m^2
Small power, including IT equipment	10–25 W/m^2

Design summer indoor conditions

Optimum temperature	20°C to 22°C
Optimum relative humidity	40% to 65%

Desirable indoor conditions in summer for exposures less than 3 hours

Outside dry bulb temp.	Inside air conditions with dewpoint constant at 14°C		
	Dry bulb	Wet bulb	Relative humidity
°C	°C	°C	%
35	27	18.5	44
32	26	18.0	46
29	25	17.8	52
27	24	17.5	51
24	23	17.2	57
21	22	17.0	57

Relation of effective temperature, to dry and wet bulb temperatures and humidity, with summer and winter comfort zones

Charts for velocities up to 0.1m/s i.e. practically still air.
For an air velocity of 0.4m/s the effective temperature decreases by 1°C.

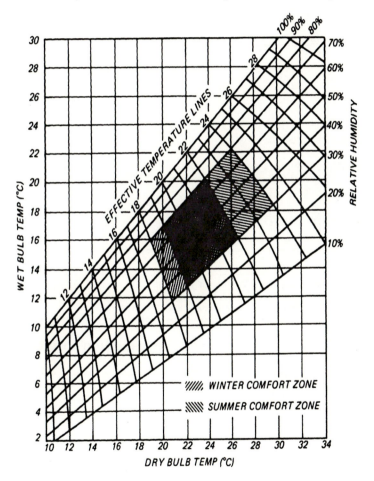

Radiation factor (F)

Proportion of radiation absorbed by wall transmitted to interior.

U-value of wall		U-value of wall	
W/m² K	F	W/m² K	F
0.15	0.006	2.5	0.10
0.25	0.01	3.0	0.12
0.5	0.02	3.5	0.14
1.0	0.04	4.0	0.16
1.5	0.06	4.5	0.18
2.0	0.08	5.0	0.20

For glass, proportion of incident radiation transmitted plus proportion absorbed and transmitted = 0.84

For translucent Filon sheeting, proportion of incident radiation transmitted plus proportion absorbed and transmitted = 0.72 for natural sheet or 0.43 for white tint sheet

Absorption coefficient (α)

Proportion of radiation falling on wall absorbed by it.

Type of surface	α
Very light surface, white stone, light cement	0.4
Medium dark surface, unpainted wood, brown stone, brick, red tile	0.7
Very dark surface, slate roofing, very dark paints	0.9

Time lag in transmission of solar radiation through walls

Type of wall	Time lag hours
150 mm concrete	3
100 mm lightweight blocks	$2\frac{1}{2}$
560 mm brick	10
75 mm concrete with 25 mm thermal insulation board	2
50 mm timber	$1\frac{1}{2}$

Transmission of radiation through shaded windows

Type of shading	Proportion transmitted
Canvas awning, plain	0.28
Canvas awning, aluminium bands	0.22
Inside shade, fully drawn	0.45
Inside shade, half drawn	0.68
Inside Venetian blind, slats at 45°, aluminium	0.58
Outside Venetian blind, slats at 45°, aluminium	0.22

Intensity of solar radiation

For latitude 45°

Solar time	Intensity of solar radiation for orientation (W/m^2)							
	NE	E	SE	S	SW	W	NW	Horizontal
5	79	75	28					6
6	281	312	164					82
7	470	612	394					284
8	441	691	539	69				492
9	290	612	577	205				663
10	104	455	539	309				791
11		237	438	382	101			864
12			287	404	287			890
13			101	382	438	237		864
14				309	539	455	104	791
15				205	577	612	290	663
16				69	539	691	441	492
17					394	612	470	284
18					164	312	281	82

Heat emitted by human body (light office or domestic work)

Still air

Air temperature	(°C)	10	12	14	16	18	20	22	24	26	28	30	32
Sensible heat	(W)	136	126	115	106	98	92	85	77	69	58	47	33
Latent heat	(W)	21	21	21	21	23	27	33	41	49	60	69	81
Total	(W)	157	147	136	127	121	119	118	118	118	118	116	114
Moisture	(g/hr)	31	31	31	31	34	40	48	60	73	88	102	120

Air velocity 1 M/S

Air temperature	(°C)	10	12	14	16	18	20	22	24	26	28	30	32
Sensible heat	(W)	152	142	131	122	112	104	97	88	81	69	55	38
Latent heat	(W)	19	19	19	19	19	20	25	32	38	49	61	77
Total	(W)	171	161	150	143	131	124	122	120	119	118	116	115
Moisture	(g/hr)	28	28	28	28	28	29	36	47	57	73	89	114

8 Heating systems

Hot water heating

Hot water carries heat through pipes from the boiler to room or space heaters.

Classification by pressure

Type	Abbreviation	Flow temp. °C	Temp. drop °C
Low pressure hot water heating	LPHW		
(a) pumped circulation		50–90	10–15
(b) gravity circulation		90	20
Medium pressure hot water	MPHW	90–120	15–35
High pressure hot water	MPHW	120–200	27–85

Classification by pipe system

One-pipe or two-pipe system ⎫
Up-feed or down-feed system ⎭ See typical schemes on page 120.

Design procedure for hot water heating system

1 Heat losses of rooms to be heated.
2 Boiler output.
3 Selection of room heating units.
4 Type, size and duty of circulating pump.
5 Pipe scheme and pipe sizes.
6 Type and size of expansion tank.

1 Heat losses

Calculated with data in section 6.

2 Boiler

$$B = H(1+X)$$

where

B = boiler rating (kW)
H = total heat loss of plant (kW)
X = margin for heating up (0.10 to 0.15)

Boilers with correct rating to be selected from manufacturers' catalogues.

HOT WATER PIPE SYSTEMS

1 PUMPED SYSTEMS

(a) OPEN EXPANSION TANK

(i) ONE-PIPE SYSTEM

(ii) TWO-PIPE SYSTEM

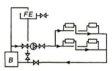

(iii) REVERSE RETURN TOTAL LENGTH OF FLOW IS THE SAME THROUGH ALL RADIATORS

(b) CLOSED EXPANSION TANK

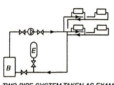

TWO-PIPE SYSTEM TAKEN AS EXAMPLE OTHER SYSTEMS ALSO POSSIBLE WITH EXPANSION TANK IN SAME RELATIVE POSITION

(c) COMBINED HEATING AND HOT WATER SYSTEM

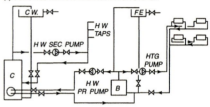

2 GRAVITY SYTEMS

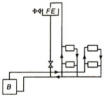

(i) TWO-PIPE UPFEED SYSTEM

(ii) TWO-PIPE DROP SYSTEM

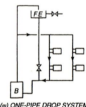

(iii) ONE-PIPE DROP SYSTEM

(iv) TWO-PIPE DROP SYTEM WITH BOILER AND RADIATORS AT SAME LEVEL

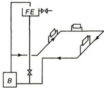

(v) ONE-PIPE RING MAIN SYSTEM

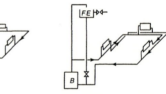

(vi) TWO-PIPE REVERSE RETURN RING MAIN SYSTEM

FE	FEED & EXPANSION TANK	C	HOT WATER CALORIFIER	⋈ VALVE		⊕ PUMP
CW	COLD WATER TANK			E CLOSED EXPANSION VESSEL		B BOILER
				⊏⊐ RADIATOR		

3 Room heaters

$$R = H\,(1+X)$$

where

R = rating of heaters in room (W)
H = heat loss of room (W)
X = margin for heating up (0.10 to 0.15)

Heaters with correct rating to be selected from manufacturers' catalogues.

4 Pump size

$$Q = \frac{H}{(h_1 - h_2)d}$$

where

Q = volume of water (m^3/s)
H = total heat loss of plant (kW)
h_1 = enthalpy of flow water (kJ/kg)
h_2 = enthalpy of return water (kJ/kg)
d = density of water at pump (kg/m^3)

For LPHW this reduces to

$$Q = \frac{H}{4.185(t_1 - t_2)}$$

where

t_1 = flow temperature (°C)
t_2 = return temperature (°C)

Pump head is chosen to give reasonable pipe sizes according to extent of system.

For LPHW 10 to 60 kN/m^2 with pipe friction resistance 80 to 250 N/m^2 per m run.

For HPHW 60 to 250 kN/m^2 with pipe friction resistance 100 to 300 N/m^2 per m run.

Gravity systems

$$p = hg(\varrho_2 - \varrho_1)$$

where

p = circulating pressure available (N/m^2)
h = height between centre of boiler and centre of radiator (m)
ϱ_1 = density of water at flow temperature (kg/m^3)
ϱ_2 = density of water at return temperature (kg/m^3)
g = acceleration of gravity = 9.81 (m/s^2)

5   Pipe sizes

$$p_T = p_1 + p_2$$
$$p_1 = il$$
$$p_2 = \sum F \frac{V^2}{2}$$

alternatively,

$$p_2 = \sum l_E$$

where

p_T = total pressure loss in system (N/m^2)
p_1 = pressure loss in pipes due to friction (N/m^2)
p_2 = pressure loss in fittings (N/m^2)
i = pipe friction resistance per length (N/m^2 per m run)
l = length of pipe (m)
F = coefficient of resistance
V = velocity of water (m/s)
ϱ = density of water (kg/m^3)
l_E = equivalent length of fitting (m)

i can be obtained from Chart 1.

Typical values of p_2/p_1

Heating installations in buildings	0.40 to 0.50
District heating mains	0.10 to 0.30
Heating mains within boiler rooms	0.70 to 0.90

6   Expansion tank

(a)   *Open tank* (For LPHW only)
Expansion of water from 7°C to 100°C = approx. 4%.
Requisite volume of expansion tank = 0.08X water contents of system. Water content for typical system is approximately 1 litre for every 1 m^2 of radiator surface.

(b)   *Closed tank*

$$V_t = V_e \frac{p_w}{p_w - p_i} \qquad V_e = V_w \frac{\varrho_1 - \varrho_2}{\varrho_2}$$

where

V_t = volume of tank (m^3)
V_e = volume by which water content expands (m^3)
V_w = volume of water in system (m^3)
p_w = pressure (absolute) of tank at working temperature (kN/m^2)
p_i = pressure (absolute) of tank when filled cold (kN/m^2)
ϱ_1 = density of water at filling temperature (kg/m^3)
ϱ_2 = density of water at working temperature (kg/m^3)

p_w to be selected so that working pressure at highest point of system corresponds to a boiling point approximately 10 K above working temperature.

p_w = working pressure at highest point + static pressure difference between highest point and tank ± pump pressure (+ or − according to position of pump).

Either p_i or V_t can be chosen independently to determine value of the other.

Approximate size of expansion tank for LPHW

Boiler rating	Tank size		Ball valve size	Cold feed size	Open vent size	Overflow size
kw	litre	BS Ref.	mm n.b.	mm n.b.	mm n.b.	mm n.b.
12	54	SCM 90	15	20	25	25
25	54	SCM 90	15	20	25	32
30	68	SCM 110	15	20	25	32
45	68	SCM 110	15	20	25	32
55	86	SCM 135	15	20	25	32
75	114	SCM 180	15	25	32	32
150	191	SCM 270	15	25	32	32
225	227	SCM 320	20	32	40	40
275	264	SCM 360	20	32	40	40
375	327	SCM 450/1	20	40	50	40
400	336	SCM 450/2	20	40	50	50
550	423	SCM 570	25	40	50	50
800	709	SCM 910	25	50	65	50
900	841	SCM 1130	25	50	65	65
1200	1227	SCM 1600	25	50	65	65

Safety valves

Safety valve setting = pressure on outlet side of pump +70 kN/m^2.

For gravity systems, safety valve setting = pressure in system +15 kN/m^2.

To prevent leakage due to shocks in system, it is recommended that the setting should be not less than 240 kN/m^2.

Valves should have clearances to allow a lift of $\frac{1}{5}$×diameter.

Safety valve sizes for water boilers

Boiler rating	Minimum clear bore of safety valves and vents
kW	mm
275	1×20
350	1×25
440	1×32
530	1×40
880	2×40
1500	80 to 150

Recommended flow temperatures for LPHW systems

Outside temperature	°C	0	2	4	7	10
Boiler flow temperature	°C	80	70	56	45	37

Resistance of fittings for LPHW pipe systems

Values of F for different fittings

Radiators	3.0	Tee, straight way	1.0
Boilers	2.5	branch	1.5
Abrupt velocity change	1.0	counter current	3.0
Cross-over	0.5		

Fitting		Nominal bore mm					
		15	20	25	32	40	50
Radiator valve:	angle	7	4	4	4	–	–
	straight	4	2	2	2	–	–
Gate valve:	screwed	1.5	0.5	0.5	0.5	0.5	0.5
	flanged	0	0	0	0	0	0
Elbow		2	2	1.5	1.5	1.0	1.0
Bend		1.5	1.5	1.0	1.0	0.5	0.5

Resistance of valves and fittings to flow of fluids in terms of equivalent length of straight pipe

Description of fitting		*Nominal diameter*										
	in	$\frac{1}{2}$	$\frac{3}{4}$	1	$1\frac{1}{4}$	$1\frac{1}{2}$	2	$2\frac{1}{2}$	3	4	5	6
	mm	15	20	25	32	40	50	65	80	100	125	150
Globe Valve	E.L. ft	13	16	26	35	40	55	65	80			
	m	4	5	8	11	12	17	20	24.5			
Angle Valve	E.L. ft	8	11	15	18	20	27	32	40			
	m	2 5	3.5	4.5	5.5	6	8.3	10	12			
Gate Valve	E.L. ft	0.3	0.5	0.5	0 5	1	1	1.5	2	2.5	3	3
	m	0 09	0.15	0.15	0.15	0.3	0.3	0.45	0.6	0.75	0 9	0.9
Elbow	E.L. ft	1	2	2	3	4	5	6	8	11	13	17
	m	0.3	0.6	0.6	0.9	1.2	1.5	1.8	2.5	3.5	4	5.2
Long Sweep Elbow	E.L. ft	1	1.5	2	2.5	3	3	5	4	6	8	10
	m	0.3	0.45	0.6	0.75	0 9	1.0	1.2	1.5	1.8	2.5	3.0
Run of Tee ⊥→	E.L. ft	1	1.5	2.5	2.5	3	3	4	5	6	8	10
	m	0.3	0.45	0.75	0.8	0.9	1.0	1.2	1 5	1.8	2.5	3.0
Run of Tee, reduced to $\frac{1}{2}$	E.L. ft	1	2	2	3	4	5	6	8	11	13	17
	m	0.3	0.5	0.7	0.9	1.2	1.5	1.8	2.5	3.5	4	5.2
Branch of Tee	E.L. ft	3.5	5	6	8	10	13	15	18	24	30	35
	m	1.1	1.5	1.8	2.5	3.0	4	4.5	5.5	7.3	9	11
Sudden Enlargement $\frac{d}{D}=\frac{1}{4}$	E.L. ft	1	2	2	3	4	5	6	8	11	13	17
	m	0.3	0.5	0.7	0.9	1.2	1.5	1.8	2.5	3.5	4	5.2
$\frac{d}{D}=\frac{1}{2}$	E.L. ft	1	1.5	2	2.5	3	3 5	4	5	7	9	11
	m	0.3	0.45	0.6	0 75	0.9	1.1	1.2	1.5	2.1	2.7	3.5
$\frac{d}{D}=\frac{3}{4}$	E.L. ft	0.3	0.5	0.5	1	1	1	1.5	2	2.3	3	3
	m	0.09	0.15	0.15	0.15	0.25	0.3	0.35	0.45	0.6	0.75	0.9	1.0
Sudden Contraction $\frac{d}{D}=\frac{1}{4}$	E.L. ft	0.8	1.0	1.2	1.5	2.0	2.5	3	4	5	6	8
	m	0.25	0.3	0.35	0.45	0.6	0.75	0.9	1.2	1.5	1.8	2.5
$\frac{d}{D}=\frac{1}{2}$	E.L. ft	0.5	0.8	1.0	1.2	1.5	2	2	3	4	5	6
	m	0.15	0.25	0.3	0.35	0.45	0.6	0.6	0.9	1.2	1.5	1.8
$\frac{d}{D}=\frac{3}{4}$	E.L. ft	0 4	0 5	0.6	1.0	1.0	1.5	1.5	2.0	2.5	3	3 5
	m	0.12	0.15	0.18	0.3	0.3	0.4	0.5	0.6	0.8	0.9	1.1
Ordinary Entrance	E.L. ft	1	1	1.5	2	2.5	3	3.5	4 5	6	8	10
	m	0 3	0.3	0.45	0.6	0.75	0.9	1 1		1.4	1.8	2.5	3 0

Circulating pressures for gravity heating

Pressure in N/m^2 per m circulating height

Return temp. (°C)	Flow temperature (°C)							
	95	90	85	80	75	70	65	60
50	257	223	190	159	129	101	74	39
55	332	200	168	136	106	97	50	24
60	209	176	143	112	82	53	26	–
65	183	150	117	87	56	27	–	–
70	156	123	90	59	28	–	–	–
75	127	94	61	30	–	–	–	–
80	98	64	31	–	–	–	–	–
85	66	32	–	–	–	–	–	–

Head in inches water gauge per foot circulating height

Return temp. (°F)	Flow temperature (°F)						
	200	190	180	170	160	150	140
120	0.324	0.277	0.230	0.187	0.145	0.104	0.068
130	0.293	0.244	0.198	0.153	0.111	0.070	0.035
140	0.258	0.210	0.163	0.118	0.077	0.036	–
150	0.221	0.172	0.126	0.081	0.040	–	–
160	0.181	0.133	0.086	0.040	–	–	–
170	0.140	0.090	0.044	–	–	–	–
180	0.096	0.046	–	–	–	–	–
190	0.048	–	–	–	–	–	–

Boiler and radiators at same level

Circulating pressure in N/m^2 for 90°C flow temperature 70°C return, return downcomers bare.

Horizontal extent of plant (m)	Horizontal distance of downcomer from main riser (m)						
	5	5–10	10–15	15–20	20–30	30–40	40–50
Up to 10	69	177	–	–	–	–	–
10–15	69	108	147	196	245	–	–
25–50	49	78	108	137	177	235	294

Head in inches water gauge for 195°F flow temperature, 160°F return, return downcomers bare.

Horizontal extent of plant (ft)	Horizontal distance of downcomer from main riser (ft)						
	16	16-32	32-48	48-64	64-96	96-125	125-160
Up to 32	0.275	0.710	–	–	–	–	–
32-82	0.275	0.430	0.600	0.800	1.00	–	–
82-164	0.200	0.310	0.430	0.550	0.710	0.950	1.180

Underfloor heating

Underfloor heating uses pipes embedded in the floor structure. The pipes carry hot water which can be provided by any of the usual sources. Heat is transferred from the pipes to the floor and the room or space is heated by low temperature radiation from the entire surface of the floor.

Proprietary systems are available from a number of manufacturers.

Pipe material:
Plastic, e.g. polyamide base thermoplastic or cross-linked polyethylene.

Pipe arrangement:
Continuous loops or modules between flow and return pipes.

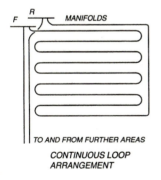

CONTINUOUS LOOP
ARRANGEMENT

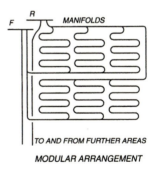

MODULAR ARRANGEMENT

Pipe sizes:
Small bore, 10 mm nb to 22 mm nb

Pipe spacing:
300 mm centres

Flow temperature:
38°C to 60°C

Temperature drop:
5 to 15 K

Floor surface temperature:
2 to 4 K above room temperature, dependent on floor finish and covering

Output:
70 to 130 W/m^2

Layout of pipes is generally determined by manufacturer or supplier of proprietary system. Installation by heating contractor in accordance with supplier's recommendations.

Individual loops or sections connected to common flow and return manifolds. A pump and mixing valve are included in the manifold assembly. Control is by mixing valve actuated in accordance with signals from a room thermostat and water flow and return temperature detectors. If required, loops from common manifolds can be controlled individually by thermostatic valves. Manifold assemblies and controls are normally included in the proprietary manufacturer's supply.

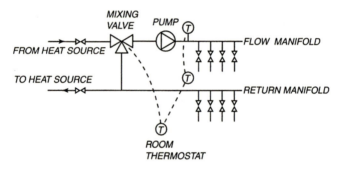

TYPICAL MANIFOLD ASSEMBLY

Advantages:
Even heating, small temperature gradient through room.

Loops can be arranged to overcome down draughts at windows.

No wall space taken up.

No high temperature surfaces, therefore safer for children, the elderly and the infirm.

No convection currents, no staining of decorations, reduced air infiltration and therefore lower heat loss.

Larger lower temperature heating surface produces comfort at lower air temperature (about 2 K), therefore reduced heating requirement.

Low flow temperature makes system suitable for condensing boilers.

Rapid response to thermostatic control.

Disadvantages:
Extra insulation needed on underside of floor.
Floor construction may have to be heavier and deeper than would otherwise be necessary.
Difficult to modify after installation.
Higher capital cost than radiator system.

Applications:
Hospitals, housing, old people's homes, sports halls, assembly halls.

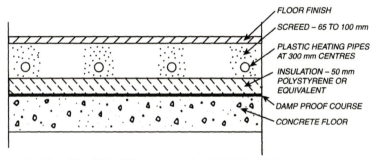

FLOOR FINISH
SCREED – 65 TO 100 mm
PLASTIC HEATING PIPES AT 300 mm CENTRES
INSULATION – 50 mm POLYSTYRENE OR EQUIVALENT
DAMP PROOF COURSE
CONCRETE FLOOR

TYPICAL CONSTRUCTION WITH CONCRETE FLOOR

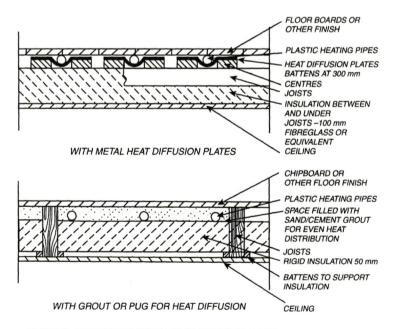

FLOOR BOARDS OR OTHER FINISH
PLASTIC HEATING PIPES
HEAT DIFFUSION PLATES
BATTENS AT 300 mm CENTRES
JOISTS
INSULATION BETWEEN AND UNDER JOISTS –100 mm FIBREGLASS OR EQUIVALENT
CEILING

WITH METAL HEAT DIFFUSION PLATES

CHIPBOARD OR OTHER FLOOR FINISH
PLASTIC HEATING PIPES
SPACE FILLED WITH SAND/CEMENT GROUT FOR EVEN HEAT DISTRIBUTION
JOISTS
RIGID INSULATION 50 mm
BATTENS TO SUPPORT INSULATION
CEILING

WITH GROUT OR PUG FOR HEAT DIFFUSION

TYPICAL CONSTRUCTIONS WITH TIMBER FLOORS

Off peak (storage) heating

Electricity is used during off peak periods to heat thermal stores from which the heat is then extracted during periods when heat is required. The stores are usually made of stone or artificial blocks having a high specific heat capacity.

Rating of unit

$$Q_1 = \frac{100 \, Q_2 T_2}{\eta \, T_1}$$

where

$Q_1 =$ input rating of unit (kW)
$Q_2 =$ heat output required (kW)
$T_1 =$ duration of input to unit (hr)
$T_2 =$ duration of heating period (hr)
$\eta =$ storage efficiency (%)

The storage efficiency allows for loss of heat from the store during the charging period. It is 90–95%.

Electrode systems

For large plants electrode boilers with water as a storage medium may be used.

Safe storage temperature is approximately 10°C below the boiling temperature at the operating pressure.

Capacity of storage vessel

$$V = \frac{1000 \, H}{4.2 \, \varrho (t_1 - t_2)}$$

where

$V =$ capacity of vessel (litre)
$H =$ heat to be stored (kJ)
$\varrho =$ density of water at storage temp. (kg/m^3)
$t_1 =$ storage temperature (°C)
$t_2 =$ return temperature (°C)

Boiler rating

$$Q = \frac{H}{3600 \, T}$$

where

$Q =$ boiler rating (kW)
$H =$ heat to be stored (kJ)
$T =$ duration of boiler opeation (hr)

High temperature H.W. heating

Flow temperature (°C)

Return temp. (°C)	210	200	195	190	185	180	175	170	165	160	155	150	145	140	135	130	125	120	115	110	105	100
75	569	538	516	493	471	449	427	405	383	351	340	318	297	275	254	232	220	190	170	147	126	105
80	547	516	494	471	449	427	405	383	361	339	318	296	275	253	232	210	198	168	145	125	104	83
85	527	496	474	451	429	407	385	363	341	319	300	278	255	233	212	190	178	148	127	105	84	63
90	506	475	453	430	408	396	374	342	320	298	277	255	234	212	191	169	158	127	106	85	63	42
95	495	454	432	409	387	365	342	321	299	277	256	234	213	191	170	148	136	106	85	63	42	21
100	464	433	411	388	366	344	322	300	278	256	235	213	192	170	149	127	115	85	64	42	21	
105	443	412	390	369	346	323	301	279	257	235	214	192	171	149	128	106	94	64	43	21		
110	422	391	369	346	324	302	280	258	236	214	193	171	150	128	107	85	73	43	22			
115	400	369	347	324	302	280	258	236	214	192	171	150	128	107	85	63	51	21				
120	379	348	327	303	281	259	237	215	193	171	150	128	107	85	64	42	21					
125	349	317	296	273	251	229	207	185	163	141	120	98	77	55	33	21						
130	337	306	284	259	239	217	195	173	151	129	108	86	65	43	22							
135	315	284	262	239	217	195	173	151	129	108	86	65	43	22								
140	294	263	241	218	196	174	151	130	108	86	65	43	22									
145	272	241	219	196	174	152	130	108	86	64	43	21										
150	251	220	198	175	153	131	109	87	65	43	22											
155	229	198	176	153	131	109	87	65	43	22												
160	208	177	155	132	110	88	66	44	22													
165	186	155	133	110	88	66	44	22														
170	164	133	111	88	66	44	22															
175	142	111	89	66	44	22																
180	122	89	67	44	22																	
185	98	67	45	22																		
190	76	45	23																			
195	53	22																				
200	31																					

Example: Flow temperature = 180°C
Return temperature = 130°C
Heat given up by 1 kg of water 217 kJ

Heat in kJ given up by 1 kg of water for various temperature drops

High temperature H.W. heating

Return temp. (°F)	Flow temperature (°F)																			
	400	390	380	370	360	350	340	330	320	310	300	290	280	270	260	250	240	230	220	210
170	237.3	226.3	215.5	204.9	194.3	183.7	173.1	162.6	152.2	141.8	131.5	121.2	110.9	100.7	90.5	80.4	70.3	60.2	50.1	40
180	227.3	216.3	205.5	194.9	184.3	173.7	163.1	152.6	142.2	131.8	121.5	111.2	100.9	90.7	80.5	70.4	60.3	50.2	40.1	30
190	217.3	206.3	195.5	184.9	174.3	163.7	153.1	142.6	132.2	121.8	111.5	101.2	90.9	80.7	70.5	60.4	50.3	40.2	30.1	20
200	207.3	196.3	185.5	174.9	164.3	153.7	143.1	132.6	122.2	111.8	101.5	91.2	80.9	70.7	60.5	50.4	40.3	30.2	20.1	10
210	197.3	186.3	175.5	164.9	154.3	143.7	133.1	122.6	112.2	101.8	91.5	81.2	70.9	60.7	50.5	40.4	30.3	20.2	10.1	
220	187.3	176.2	165.4	154.8	144.2	133.6	123.0	112.5	102.1	91.7	81.4	71.1	60.8	50.6	40.4	30.3	20.2	10.1		
230	177.1	166.1	155.3	144.7	134.1	123.6	112.9	102.4	92.0	81.6	71.3	61.0	50.7	40.5	30.4	20.2	10.1			
240	167.0	156.0	145.2	134.6	124.0	113.4	102.8	92.3	81.9	71.5	61.2	50.9	40.6	30.4	20.3	10.1				
250	156.3	145.9	135.1	124.5	113.9	103.3	92.7	82.2	71.8	61.4	51.1	40.8	30.5	20.3	10.1					
260	146.7	135.7	124.9	114.3	103.7	93.1	82.5	72.0	61.6	51.2	40.9	30.6	20.3	10.2						
270	136.6	125.6	114.8	104.2	93.6	83.0	72.4	61.9	51.5	41.1	30.8	20.5	10.2							
280	126.4	115.4	104.6	94.0	83.4	72.8	62.1	51.9	41.3	31.0	20.6	10.3								
290	116.1	105.1	94.3	83.7	73.1	62.5	51.9	41.4	31.0	20.6	10.3									
300	105.8	94.3	84.0	73.4	62.8	52.2	41.6	31.1	20.7	10.8										
310	95.5	84.5	73.7	63.1	52.5	41.9	31.3	20.8	10.4											
320	85.0	74.5	63.2	51.7	41.1	30.5	20.9	10.4												
330	74.7	63.7	52.9	42.3	31.7	21.1	10.5													
340	64.1	53.1	42.3	31.8	21.2	10.6														
350	53.6	42.6	31.8	21.2	10.6															
360	42.0	31.0	21.4	10.6																
370	32.4	21.0	10.6																	
380	21.8	10.8																		
390	11.0																			

Example: Flow temperature = 350°F
Return temperature = 240°F
Heat given up by 1 lb of water is 113.4 Btu per lb

Heat in Btu given up by 1 lb of water for various temperature drops

Heat pumps

The heat pump is a common refrigeration unit arranged in such a way that it can be used for both cooling and heating, or for heating only. The initial cost of the installation is high, and savings and advantages are achieved mainly when heating and cooling are required in winter and summer respectively.

Operation of the heat pump:
Referring to the scheme drawing below, the heat pump consists of the following parts:

Compressor, with driving motor, for raising the pressure and temperature of the refrigerant vapour.

Condenser, for extracting heat from the refrigerant.

Receiver (storage tank) to hold the liquid refrigerant in the high pressure side before it passes the expansion valve.

Expansion valve, for causing expansion of the refrigerant and for lowering the pressure from the high pressure to the low pressure side of the system.

Evaporator, in which heat is absorbed by the refrigerant from some source. Water, earth or air can be used as the source of heat.

A commercial refrigeration unit and heat pump consist of the same units and the same plant can be used either for cooling or heating.

The changing of the system from cooling to heating can be carried out by either of the following methods

(a) Leave the flow of the refrigerant unchanged and change the circuit of the heat source and the medium to be heated.

(b) Leave the heat source and the medium to be heated unchanged and reverse the flow of the refrigerant by a suitable pipe and valve scheme.

Schemes for a heat pump indicating suitable temperatures when used for cooling and heating are shown, the data being chosen for the purpose of illustration only.

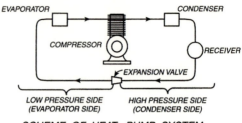

SCHEME OF HEAT PUMP SYSTEM

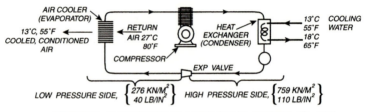

COOLING CYCLE OF THE HEAT PUMP (WATER TO AIR)

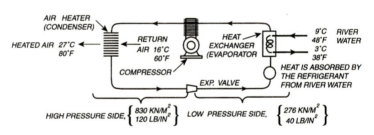

HEATING CYCLE OF THE HEAT PUMP (WATER TO AIR)

9 Steam systems

Steam heating

Steam carries heat through pipes from the boiler to room or space heaters.

This is now seldom used as a method of space heating. This section is included for reference when old systems have to be examined or altered, and for design of systems in industrial premises where steam is available and steam-to-water calorifiers cannot be justified. Steam is also used for process heating in industry and the data in this section can also be used for pipe sizing.

Classification of steam heating systems

1 By pressure
   (a) High pressure steam heating system
   (b) Low pressure steam heating system
       Up to about 3 lb/in^2 or 20 kN/m^2
   (c) Vacuum system

2 By method of returning condensate
   (a) Gravity system
   (b) Mechanical system

3 By pipe scheme
   (a) One-pipe or two-pipe system
   (b) Up-feed or down-feed system

(See illustrations on page 136.)

Steam heating systems

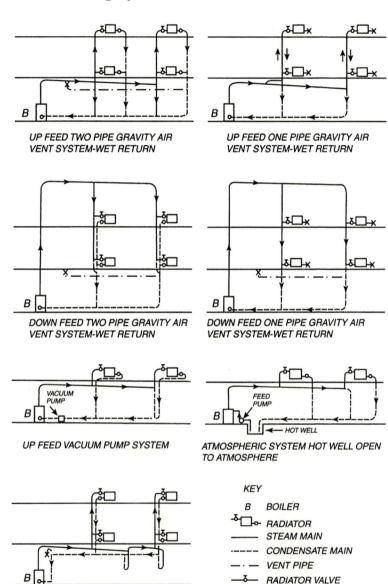

UP FEED TWO PIPE GRAVITY AIR
VENT SYSTEM-WET RETURN

UP FEED ONE PIPE GRAVITY AIR
VENT SYSTEM-WET RETURN

DOWN FEED TWO PIPE GRAVITY AIR
VENT SYSTEM-WET RETURN

DOWN FEED ONE PIPE GRAVITY AIR
VENT SYSTEM-WET RETURN

VACUUM
PUMP

UP FEED VACUUM PUMP SYSTEM

FEED
PUMP

HOT WELL

ATMOSPHERIC SYSTEM HOT WELL OPEN
TO ATMOSPHERE

UP FEED TWO PIPE GRAVITY SYSTEM-
DRY RETURN

KEY

B   BOILER

     RADIATOR

——— STEAM MAIN

----- CONDENSATE MAIN

— · — VENT PIPE

     RADIATOR VALVE

     STEAM TRAP

—× VENT

Vacuum differential heating system

In vacuum steam heating systems, a partial vacuum is maintained in the return line by means of a vacuum pump. The vacuum maintained is approx. 3-10 in mercury = approx. 75-250 mm mercury.

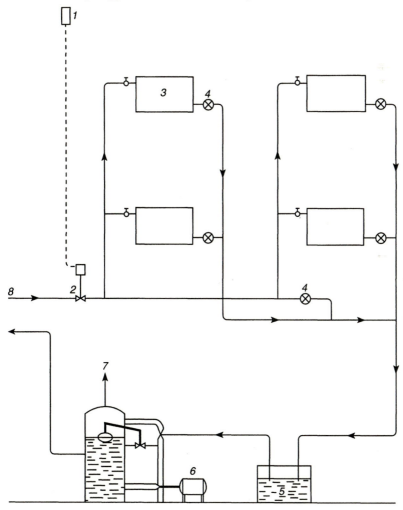

1 OUTSIDE THERMOSTAT   5 CONDENSE RECEIVER
2 CONTROL VALVE   6 VACUUM PUMP
3 RADIATORS   7 VENT
4 STEAM TRAPS   8 STEAM SUPPLY

Capacities of condensate pipes in watts

| Nominal pipe size | | Wet main | Dry main with gradient | | | Vent pipes |
in	mm		1 in 200	1 in 600	Vertical	
$\frac{1}{2}$	15	30 000	10 000	6 000	10 000	12 000
$\frac{3}{4}$	20	70 000	30 000	18 000	30 000	47 000
1	25	120 000	50 000	34 000	50 000	94 000
$1\frac{1}{4}$	32	300 000	120 000	80 000	120 000	211 000
$1\frac{1}{2}$	40	420 000	176 000	117 000	176 000	293 000
2	50	760 000	350 000	225 000	350 000	530 000
$2\frac{1}{2}$	65	1 900 000	800 000	510 000	800 000	1 200 000
3	80	2 700 000	1 200 000	740 000	1 200 000	1 870 000

Capacities of condensate pipes in Btu/hr

| Nominal pipe size | | Wet main | Dry main with gradient | | | Vent pipes |
in	mm		$\frac{3}{16}$ in per yd	$\frac{1}{16}$ in per yd	Vertical	
$\frac{1}{2}$	15	100 000	40 000	24 000	40 000	40 000
$\frac{3}{4}$	20	240 000	108 000	68 000	108 000	160 000
1	25	400 000	192 000	120 000	192 000	320 000
$1\frac{1}{4}$	32	1 000 000	440 000	280 000	440 000	720 000
$1\frac{1}{2}$	40	1 440 000	600 000	400 000	600 000	1 000 000
2	50	2 600 000	1 120 000	700 000	1 120 000	1 800 000
$2\frac{1}{2}$	65	6 400 000	2 800 000	1 760 000	2 800 000	4 000 000
3	80	9 600 000	4 000 000	2 520 000	4 000 000	6 400 000

Safety valves for steam heating
(Working pressure = 70 kN/m^2)

| Output Watts | Minimum clear bore | | Output Btu/hr | Minimum clear bore | |
	in	mm		in	mm
24 000	$\frac{3}{4}$	20	80 000	$\frac{3}{4}$	20
44 000	1	25	150 000	1	25
73 000	$1\frac{1}{4}$	32	250 000	$1\frac{1}{4}$	32
100 000	$1\frac{1}{2}$	40	350 000	$1\frac{1}{2}$	40
230 000	2	50	800 000	2	50
275 000	$2\frac{1}{2}$	65	950 000	$2\frac{1}{2}$	65
440 000	Two 2	Two 50	1 500 000	Two 2	Two 50

Suction lift of boiler feed pumps for various water temperatures

Temperature of feed water °F	Maximum suction lift (ft)	Minimum pressure head (ft)	Temperature of feed water (°C)	Maximum suction lift (m)	Minimum pressure head (m)
130	10		55	3	
150	2		65	2	
170	7		77	0.6	
175	0	0	80	0	0
190		5	87.5		1.5
200		10	95		3.5
210		15	99		4.5
212		17	100		5.0

Quantities of flash steam

Condensate Absolute pressure (kN/m^2)	Temperature (°C)	Percentage of condensate flashed off at reduction of pressure to kN/m^2 absolute					
		400	260	170	101.33	65	35
1500	198.3	11.3	14.0	16.4	18.9	20.4	23.2
1150	186.0	8.7	11.5	13.9	16.5	18.4	20.9
800	170.4	5.5	8.2	10.8	13.4	15.4	17.9
650	162.0	3.7	6.5	9.1	11.8	13.7	16.3
500	151.8	1.6	4.6	7.1	9.8	11.8	14.4
400	143.6	–	3.0	5.5	8.3	10.3	12.9
260	128.7	–	–	2.6	5.4	7.5	10.2
170	115.2	–	–	–	2.8	5.0	7.7
101.33	100	–	–	–	–	2.2	4.9

Condensate Gauge pressure (lb/in^2)	Temperature (°F)	Percentage of condensate flashed off at reduction of pressure to lb/in^2 gauge or in Hg vacuum					
		40	20	10	0	10 in	20 in
200	388	11.5	14.3	16.2	18.8	20.5	23.2
150	366	9.0	11.8	13.0	16.4	18.2	20.9
100	338	5.8	8.6	10.6	13.3	15.1	17.9
80	324	4.2	7.1	9.1	11.9	13.7	16.5
60	308	2.3	5.2	7.3	10.0	11.8	14.7
40	287	–	3.0	5.0	7.8	9.7	12.6
20	259	–	–	2.1	5.0	6.8	9.8
10	240	–	–	–	2.9	4.8	7.8
0	212	–	–	–	–	1.9	5.0

CHART 1. PIPE SIZES FOR HOT WATER HEATING

$$\left(\frac{1}{1,000} \text{ in. PER ft.}\right)$$

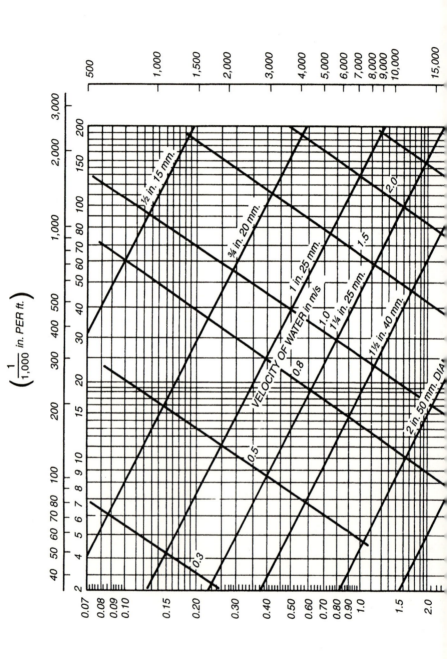

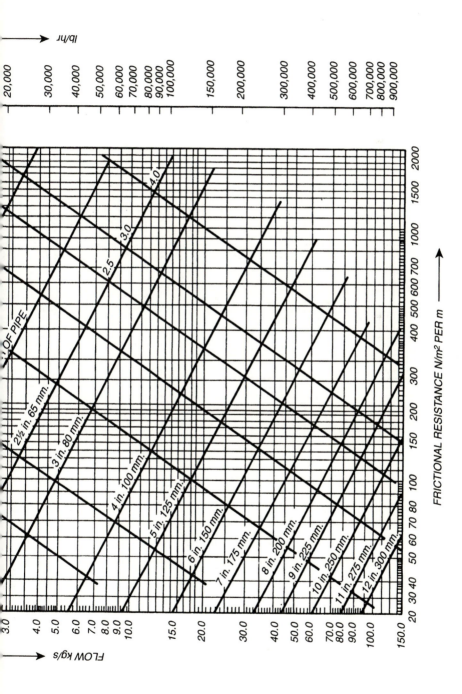

CHART 1a. PIPE SIZES FOR HOT WATER HEATING

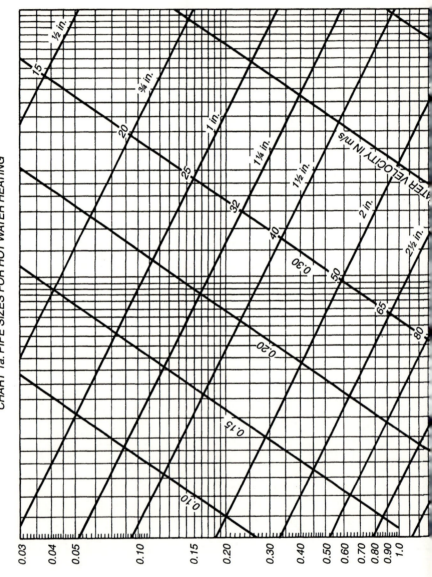

WATER VELOCITY IN m/s

WATER IN kg/s

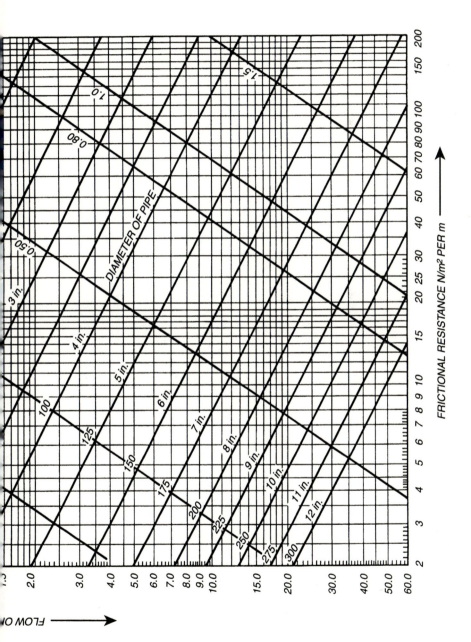

FRICTIONAL RESISTANCE N/m² PER m

FLOW O

CHART 2. PIPE SIZES FOR LOW PRESSURE STEAM HEATING

$$\left(\frac{1}{1,000} \; lb/in^2 \; PER \; ft \; RUN\right)$$

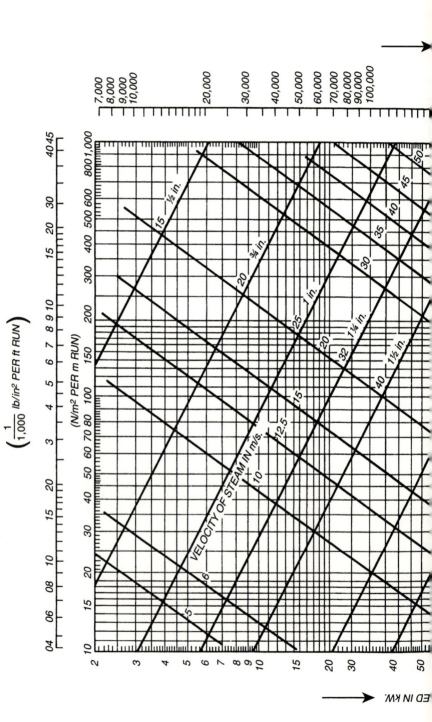

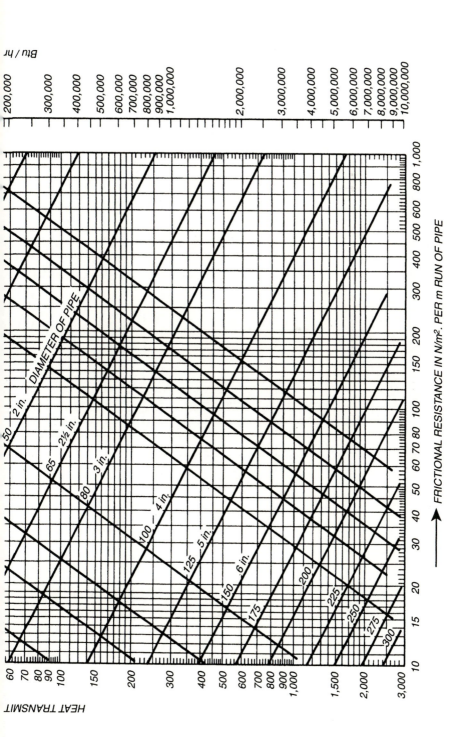

Btu / hr

200,000
300,000
400,000
500,000
600,000
700,000
800,000
900,000
1,000,000
2,000,000
3,000,000
4,000,000
5,000,000
6,000,000
7,000,000
8,000,000
9,000,000
10,000,000

FRICTIONAL RESISTANCE IN N/m². PER m RUN OF PIPE

DIAMETER OF PIPE

50 — 2 in.
65 — 2½ in.
80 — 3 in.
100 — 4 in.
125 — 5 in.
150 — 6 in.
175
200
225
250
275
300

HEAT TRANSMIT

60
70
80
90
100
150
200
300
400
500
600
700
800
900
1,000
1,500
2,000
3,000

10 15 20 30 40 50 60 70 80 100 150 200 300 400 500 600 800 1,000

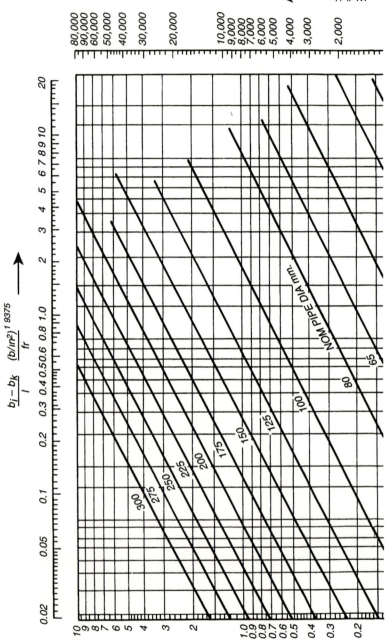

CHART 3. PIPE SIZES FOR HIGH PRESSURE STEAM

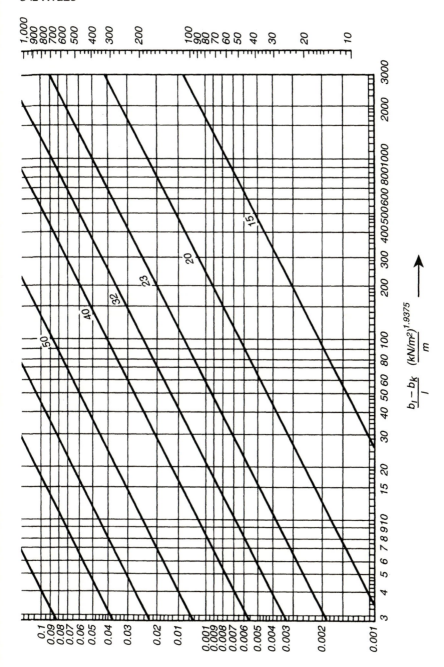

$$\frac{b_l - b_K}{l} \quad \frac{(kN/m^2)^{1.9375}}{m}$$

Scheme of flash steam recovery

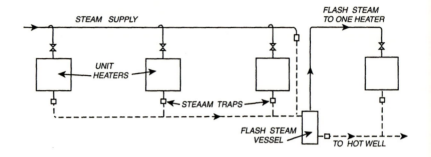

Sizing steam mains

The *Available Pressure Drop* is the difference between the initial or boiler pressure and the required final pressure at the end of the line.

$$p = p_j - p_k$$

The available pressure drop is used to overcome friction in pipes and pressure losses in fittings.

$$p_t = p_1 + p_2$$

For *low pressure steam*

$$p_1 = p_a l$$

$$p_2 = \sum F \frac{v^2 \varrho}{2}$$

Alternatively

$$p_2 = p_a l_e$$

and

$$p_t = p_a (l + l_e)$$

p_a can be read from Chart 2 for given steam flow and pipe size.

For *high pressure steam* Chart 3 can be used. In this the auxiliary value b_x is used in place of the pressure drop per unit length.

$$b_x = b_j - b_k$$
$$b_j = p_j^{1\,9375}$$
$$b_k = p_k^{1\,9375}$$

In the above formula

p_t = total pressure drop in system (N/m^2)
p_j = initial or boiler pressure (N/m^2)
p_k = final pressure (N/m^2)
p_1 = pressure loss in pipes due to friction (N/m^2)
p_2 = pressure loss in fittings (N/m^2)
p_a = pipe friction resistance per length (N/m^2)
F = coefficient of resistance
v = steam velocity (m/s)
ϱ = density of steam (kg/m^3)
l = length of pipe (m)
l_e = equivalent length of fitting (m)

Ratio p_2/p_1 is generally about 0.33.

Total pressure drop is generally about 6 per cent of initial pressure per 100 m of pipe system.

Typical steam velocities

Exhaust steam	20–30 m/s	(70–100 ft/s)
Saturated steam	30–40 m/s	(100–130 ft/s)
Superheated steam	40–60 m/s	(130–200 ft/s)

Values of F for fittings

	Nom bore					
	$\frac{1}{2}$ in	$\frac{3}{4}$ in	1 in	$1\frac{1}{4}$ in	$1\frac{1}{2}$ in	2 in
Fitting	15 mm	20 mm	25 mm	32 mm	40 mm	50 mm
Radiator	1.5	1.5	1.5	1.5	1.5	1.5
Abrupt velocity change	1.0	1.0	1.0	1.0	1.0	1.0
Cross over	0.5	0.5	0.5	0.5	0.5	0.5
Angle valve	9	9	9	9		
Globe valve	15	17	19	30		
Angle cock	7	4	4	4		
Straight cock	4	2	2	2		
Gate valve	1.5	0.5	0.5	0.5	0.5	0.5
Damper	3.5	2	2	1.5	1.5	1
Elbow	2	2	1.5	1.5	1	1
Long sweep elbow	1.5	1.5	1	1	0.5	0.5
Short radius bend	2	2	2	2	2	2
Long radius bend	1	1	1	1	1	1
Tee straight	1	1	1	1	1	1
branch	1.5	1.5	1.5	1.5	1.5	1.5
counter current	3.0	3.0	3.0	3.0	3.0	3.0
double branch	1.5	1.5	1.5	1.5	1.5	1.5

High pressure steam pipes
Resistance of valves and fittings to flow of steam
Expressed as an equivalent length of straight pipe

| Nom bore of pipe | | Bends of standard radius | | | | Barrel of tee | | | | Branch of tee | | Valves | | | | | | Lyre expansion bends | |
| | | 90° | | 45° | | Plain | | Reduced 25% | | | | Through | | Angle | | Globe | | | |
in	mm	ft	m	ft	m	ft	m	ft	m	ft	m	ft	m	ft	m	ft	m	ft	m
1	25	0.5	0.15	0.4	0.12	0.5	0.15	0.7	0.21	2.2	0.67	0.4	0.12	1.5	0.46	3.3	1.0	2.2	0.67
1¼	32	0.7	0.21	0.5	0.15	0.7	0.21	0.9	0.27	2.9	0.89	0.5	0.15	2.0	0.61	4.3	1.3	2.9	0.88
1½	40	0.9	0.27	0.7	0.21	0.9	0.27	1.1	0.33	3.6	1.1	0.7	0.21	2.4	0.73	5.4	1.6	3.6	1.1
2	50	1.3	0.40	1.0	0.30	1.3	0.40	1.6	0.49	5.1	1.6	1.3	0.40	3.4	1.0	7.6	2.3	5.1	1.6
2½	65	1.6	0.49	1.2	0.37	1.6	0.49	2.1	0.64	6.6	2.0	1.6	0.49	4.5	1.4	10.0	3.0	6.6	2.0
3	80	2.1	0.64	1.6	0.49	2.1	0.64	2.6	0.80	8.3	2.5	2.1	0.64	5.6	1.7	12.0	3.7	8.3	2.5
4	100	2.9	0.88	2.2	0.67	2.9	0.88	3.7	1.1	12.0	3.7	2.2	0.67	7.9	2.4	18.0	5.5	12.0	3.7
5	125	3.8	1.2	2.9	0.88	3.8	1.2	4.8	1.5	15.0	4.6	2.9	0.89	10.0	3.0	23.0	7.0	15.0	4.6
6	150	4.7	1.4	3.6	1.1	4.7	1.4	6.0	1.8	19.0	5.8	3.6	1.1	13.0	4.0	29.0	8.8	19.0	5.8
7	175	5.7	1.7	4.3	1.3	5.7	1.7	7.2	2.2	23.0	7.0	4.3	1.3	15.0	4.6	34.0	10	23.0	7.0
8	200	6.7	2.0	5.0	1.5	7.6	2.0	8.5	2.6	27.0	8.2	5.0	1.5	18.0	5.5	40.0	12	27.0	8.2
9	225	7.7	2.3	5.8	1.8	7.7	2.3	9.8	3.0	31.0	10	5.8	1.7	21.0	6.4	46.0	14	31.0	9.5
10	250	8.7	2.7	6.6	2.0	8.7	2.7	11.0	3.4	35.0	11	6.6	1.8	24.0	7.3	53.0	16	35.0	11

10 Domestic services

Domestic hot water supply

Classification
Direct System. Secondary water is heated by direct mixing with boiler water in a hot water cylinder.

Indirect system. Secondary water is heated by indirect heating by primary water from boiler in an indirect cylinder or calorifier.

Design procedure for domestic hot water

1 Determination of demand (quantity and temperature).
2 Selection of type, capacity and heating surface of calorifier.
3 Selection of boiler.
4 Pipe scheme and pipe sizes.

1 Demand
Hot water is normally stored and supplied at 60°C. For canteens and large kitchens it may be required at 65°C. Where lower temperatures are necessary for safety (e.g. nursery schools, centres for handicapped) it may be stored and supplied at a lower temperature (usually 40°-50°C) or stored and supplied at a higher temperature and reduced by mixing with cold water in a blender at the point of draw off.

Quantity is determined either according to number of occupants or according to number of fittings.

2 Calorifier

$$H = \frac{4.2V(\theta_2 - \theta_1)}{3600\,t}$$

where

H = heating capacity (kW)
V = volume stored (litre)
θ_1 = temperature of cold feed water (°C)
θ_2 = temperature of hot water (°C)
t = time in which contents are to be raised from θ_1 to θ_2 (hr)

For instantaneous heating (non-storage calorifier or direct heater).

$$H = 4.2\,v(\theta_2 - \theta_1)$$

where

v = demand in litre/s.

Storage systems are usually designed for $t = 1$ hr or 2 hr. A shorter warming up time enables the volume of the calorifier to be reduced but may require a higher rate of heating.

Heating Surface

$$A = \frac{1000\,H}{k\,\theta_m}$$

$$\theta_m = \frac{\theta_f - \theta_r + \theta_2 - \theta_1}{2.3 \log_{10} \theta_f - \theta_1 / \theta_r - \theta_2}$$

where

A = heating surface of calorifier (m^2)
H = rate of heating (kW)
k = heat transmission coefficient (W/m^2 K)
θ_m = logarithmic mean temperature difference (K)
θ_f = primary flow temperature (°C)
θ_r = primary return temperature (°C)
θ_1 = secondary inlet temperature (°C)
θ_2 = secondary final temperature (°C)

3   Boiler

Boiler rating = Heating capacity of calorifier
Boiler with correct rating to be selected from manufacturers' catalogues.

4   Pipe sizes

Pipes can be sized as for hot water heating systems (see section 8, page 122).

Volume flow through pipes to draw offs is determined by maximum demand. Volume flow through return pipes of circulating system is made sufficient to keep temperature drop between flow and return connections of calorifier down to about 5 K.

For most schemes pipe sizing table on page 155 is satisfactory.

Pump duties can be determined as for hot water heating systems (see section 8, page 121).

Precautions against legionellosis

Infection is caused by inhalation of airborne droplets containing viable legionella. Sources can be hot and cold water services, cooling towers, humidifiers, air washers.

1   Pipe lengths and dead legs as short as possible.

Design temperatures to be maintained by adequate insulation, including insulation of cold pipes to prevent rise of temperature.

Adequate access for regular cleaning.

Water should not stand for long periods in conditions where its temperature may rise above 20°C.

2   Hot water storage at 60°C with at last 50°C attained at outlets after one minute of running. Stratification in calorifiers to be avoided.

Cold water storage and distribution at 20°C or below.

There is a risk of scalding at 43°C and above. Thermostatic mixing valves to be used as close as possible to outlets.

3. Alternatives to temperature control are water treatment by ionisation with copper and silver, dosing with chlorine dioxide, and biocidal treatment with ozone or ultra violet light.

Domestic hot water schemes

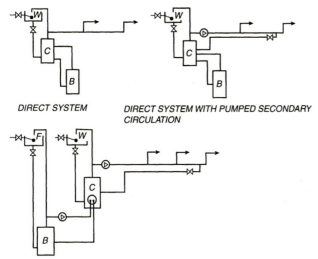

DIRECT SYSTEM

DIRECT SYSTEM WITH PUMPED SECONDARY CIRCULATION

INDIRECT SYSTEM WITH PUMPED PRIMARY AND PUMPED SECONDARY

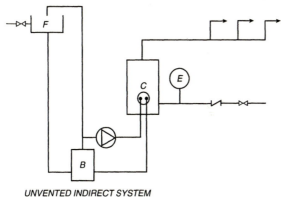

UNVENTED INDIRECT SYSTEM

LEGEND

W	COLD WATER TANK	E	CLOSED EXPANSION VESSEL
F	PRIMARY FEED AND EXPANSION TANK	B	BOILER
C	DIRECT CYLINDER	PUMP	
		VALVE	
		DRAW-OFF TAP	
C	INDIRECT CALORIFIER	NON-RETURN VALVE	

Hot water consumption per fitting

Fitting	Consumption litre/hr	gal/hr	Fitting	Consumption litre/hr	gal/hr
Basin (private)	14	3	Sink	45–90	10–20
Basin (public)	45	10	Bath	90–180	20–40
Shower	180	40			

Hot water consumption per occupant

Type of building	Consumption per occupant litre/day	gal/day	Peak demand per occupant litre/hr	gal/hr	Storage per occupant litre	gal
Factories (no process)	22–45	5–10	9	2	5	1
Hospitals, general	160	35	30	7	27	6
mental	110	25	22	5	27	6
Hostels	120	26	50	11	30	7
Hotels	130–230	28–50	50	11	30	7
Houses and flats	45–160	10–35	50	11	30	7
Offices	22	5	9	2	5	1
Schools, boarding	115	25	30	7	25	5
day	10	2	9	2	5	1

Contents of fittings

Fitting	Contents litre	gal
Basin, normal	4	0.8
Basin, full	9	2
Sink, normal	18	4
Sink, full	30	6.5
Bath	100–135	22–30

Flow rates

Fitting	Flow rate litre/s	gal/min
Basin	0.08	1
Sink	0.15	2
Bath	0.15	2
Shower	0.09–0.12	1.2–1.6

Maximum dead leg of hot water pipe without circulation

Pipe size Steel	Copper	Length m
15	15	12
20	22	8
25	28	3

Pipe sizes for domestic cold and hot water service

Nominal bore of pipe			Maximum number of draw offs served		
			Flow pipes		
in	Steel pipe mm	Copper pipe mm	Head up to 20 m (70 ft)	Head over 20 m (70 ft)	Return pipes
$\frac{1}{2}$	15	15	1	1 to 2	1 to 8
$\frac{3}{4}$	20	22	2 to 4	3 to 9	9 to 29
1	25	28	5 to 8	10 to 19	30 to 66
$1\frac{1}{4}$	32	35	9 to 24	20 to 49	67 to 169
$1\frac{1}{2}$	40	42	25 to 49	50 to 79	170 to 350
2	50	54	50 to 99	80 to 153	–
$2\frac{1}{2}$	65	67	100 to 200	154 to 300	–

For the purpose of this table, basins, sinks, showers count as one draw off, baths count as two draw offs.

Cold water storage per occupant

Type of building	Storage per occupant		Type of building	Storage per occupant	
	litres	gal		litres	gal
Factories (no process)	10	2	Offices with canteen	45	10
Hospitals			without canteen	35	8
per bed	150	33	Restaurant,		
per staff on duty	45	10	per meal	7	1.5
Hostels	90	20	Schools		
Hotels	150	33	boarding	90	20
Houses and flats	135	30	day	30	7

Cold water storage per fitting

Type of fitting	Storage per unit		Type of fitting	Storage per unit	
	litres	gal		litres	gal
Shower	450–900	100–200	Sink	90	20
Bath	900	200	Urinal	180	40
W.C.	180	40	Garden		
Basin	90	20	watering tap	180	40

Temperature drop in bare pipes

Flow of water kg/s	Temperature drop K/m for size of pipe								
	15 mm	20 mm	25 mm	32 mm	40 mm	50 mm	65 mm	80 mm	100 mm
0.010	1.03	1.37	1.49	1.83	2.06	2.52	2.88	3.44	4.35
0.012	0.86	1.14	1.24	1.54	1.72	2.10	2.40	2.87	3.63
0.014	0.74	0.98	1.06	1.31	1.45	1.80	2.06	2.43	3.11
0.016	0.65	0.86	0.93	1.14	1.29	1.57	1.80	2.14	2.72
0.018	0.57	0.76	0.83	1.02	1.14	1.40	1.60	1.91	2.42
0.020	0.52	0.69	0.74	0.92	1.03	1.26	1.44	1.77	2.08
0.025	0.41	0.55	0.60	0.72	0.82	1.01	1.16	1.37	1.78
0.030	0.34	0.45	0.50	0.61	0.69	0.84	0.96	1.15	1.45
0.035	0.29	0.39	0.43	0.52	0.59	0.72	0.82	0.98	1.24
0.040	0.26	0.34	0.39	0.46	0.52	0.63	0.72	0.86	1.09
0.045	0.23	0.30	0.33	0.41	0.46	0.56	0.64	0.76	0.97
0.050	0.21	0.27	0.30	0.37	0.41	0.50	0.57	0.69	0.87
0.060	0.17	0.23	0.25	0.31	0.34	0.42	0.48	0.57	0.76
0.070	0.15	0.20	0.21	0.26	0.29	0.36	0.41	0.49	0.62
0.080	0.13	0.17	0.19	0.23	0.26	0.32	0.36	0.47	0.54
0.090	0.11	0.15	0.17	0.20	0.23	0.27	0.32	0.43	0.48
0.100	0.10	0.14	0.15	0.18	0.21	0.25	0.29	0.34	0.44

Flow of water lb/hr	Temperature drop °F/ft for size of pipe								
	$\frac{1}{2}$ in	$\frac{3}{4}$ in	1 in	$1\frac{1}{4}$ in	$1\frac{1}{2}$ in	2 in	$2\frac{1}{2}$ in	3 in	4 in
100	0.45	0.60	0.65	0.80	0.90	1.10	1.25	1.50	1.90
120	0.38	0.50	0.54	0.62	0.75	0.92	1.04	1.21	1.58
140	0.32	0.43	0.46	0.57	0.64	0.79	0.89	1.07	1.36
160	0.28	0.37	0.41	0.50	0.56	0.69	0.78	0.94	1.19
180	0.25	0.33	0.36	0.44	0.50	0.61	0.69	0.83	1.06
200	0.22	0.30	0.33	0.40	0.45	0.55	0.63	0.75	0.95
250	0.18	0.24	0.26	0.32	0.36	0.44	0.50	0.60	0.76
300	0.15	0.20	0.22	0.27	0.30	0.37	0.42	0.50	0.63
350	0.13	0.17	0.19	0.23	0.26	0.31	0.36	0.43	0.54
400	0.11	0.15	0.17	0.20	0.23	0.28	0.32	0.38	0.48
450	0.10	0.13	0.14	0.18	0.20	0.24	0.28	0.33	0.42
500	0.09	0.12	0.13	0.16	0.18	0.22	0.25	0.30	0.38
600	0.075	0.10	0.11	0.14	0.15	0.19	0.21	0.25	0.37
700	0.065	0.085	0.095	0.13	0.13	0.16	0.18	0.22	0.27
800	0.055	0.075	0.083	0.10	0.11	0.14	0.16	0.19	0.24
900	0.050	0.066	0.070	0.089	0.10	0.12	0.14	0.17	0.21
1000	0.045	0.060	0.065	0.080	0.090	0.11	0.13	0.15	0.19

Cold water storage systems for tall buildings

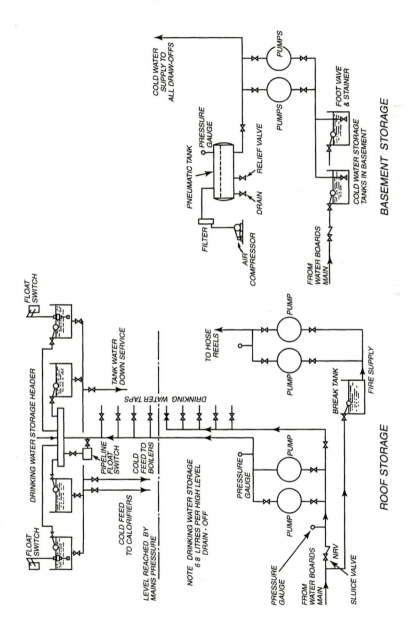

Fire service

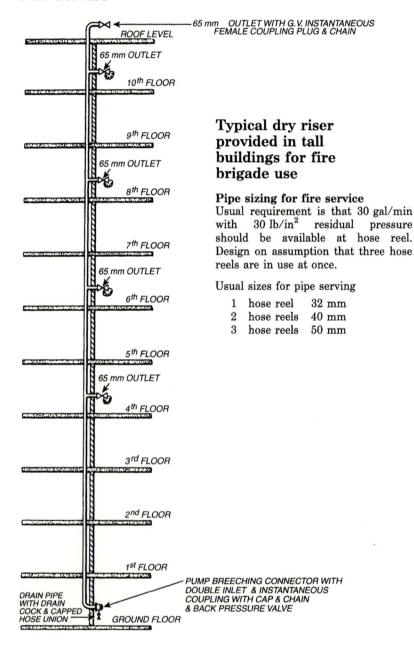

65 mm   OUTLET WITH G.V. INSTANTANEOUS
FEMALE COUPLING PLUG & CHAIN

ROOF LEVEL

65 mm OUTLET

10th FLOOR

9th FLOOR

65 mm OUTLET

8th FLOOR

7th FLOOR

65 mm OUTLET

6th FLOOR

5th FLOOR

65 mm OUTLET

4th FLOOR

3rd FLOOR

2nd FLOOR

1st FLOOR

DRAIN PIPE
WITH DRAIN
COCK & CAPPED
HOSE UNION

GROUND FLOOR

PUMP BREECHING CONNECTOR WITH
DOUBLE INLET & INSTANTANEOUS
COUPLING WITH CAP & CHAIN
& BACK PRESSURE VALVE

Typical dry riser provided in tall buildings for fire brigade use

Pipe sizing for fire service

Usual requirement is that 30 gal/min with 30 lb/in^2 residual pressure should be available at hose reel. Design on assumption that three hose reels are in use at once.

Usual sizes for pipe serving

1	hose reel	32 mm
2	hose reels	40 mm
3	hose reels	50 mm

Gas supply. Gas consumption of equipment (natural gas)

	ft^3/h	m^3/s	litre/s	Heat dissipated kW
10 gal boiling pan	45	350×10^{-6}	0.35	13
20 gal boiling pan	60	475×10^{-6}	0.48	18
30 gal boiling pan	75	600×10^{-6}	0.60	22
40 gal boiling pan	90	700×10^{-6}	0.70	26
4 ft hot cupboard	48	375×10^{-6}	0.38	14
6 ft hot cupboard	54	425×10^{-6}	0.43	16
Steaming oven	40 to 50	300 to 400×10^{-6}	0.30 to 0.40	11 to 15
Double steaming oven	100	800×10^{-6}	0.80	30
2-tier roasting oven	50	400×10^{-6}	0.40	15
Double oven range	400	3200×10^{-6}	3.2	115
Roasting oven	30	240×10^{-6}	0.24	9
Gas cooker	75	600×10^{-6}	0.60	20
Hot cupboard	17	140×10^{-6}	0.14	5
Drying cupboard	5	40×10^{-6}	0.04	1.5
Gas iron heater	5	40×10^{-6}	0.04	1.5
Washing machine	20	150×10^{-6}	0.15	6
Wash boiler	30 to 50	230 to 400×10^{-6}	0.23 to 0.40	8 to 15
Bunsen burner	3	20×10^{-6}	0.02	1
Bunsen burner, full on	10	80×10^{-6}	0.08	3
Glue kettle	10	80×10^{-6}	0.08	3
Forge	15	115×10^{-6}	0.12	4
Brazing hearth	30	230×10^{-6}	0.23	9

Flow of gas in steel tubes

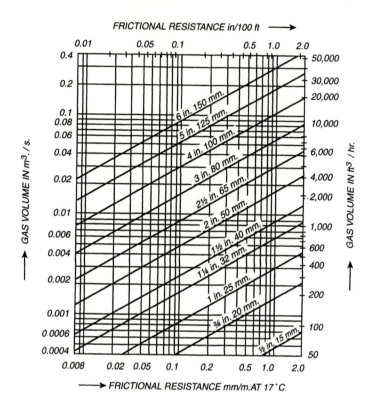

11 Ventilation

Ventilation

Classification by distribution

Central system. A central plant supplies air to the whole building. There can also be a central extract system.

Unit system. Each room or area of the building has its own ventilating unit.

Classification by function

Split system of heating and ventilating. Heat losses through the fabric of the building are supplied by a radiator heating system and the ventilation delivers air at room temperature.

Combined system. A central ventilation plant supplies air above or below room temperature so that in cooling or heating to room temperature it provides the required heating or cooling as well as ventilation.

Schemes of air distribution

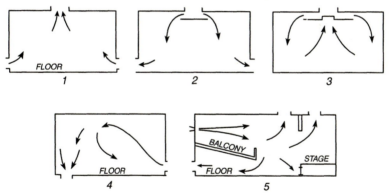

DIAGRAMMATIC VIEWS (IN ELEVATION) SHOWING HOW VARIOUS SYTEMS OF AIR DISTRIBUTION ARE APPLIED IN BUILDINGS.

1 Upward flow system
2 Downward flow system
3 High-level supply and return system
4 Low-level supply and return system
5 Ejector system

Design procedure for ventilating system

1   Heating or cooling load, including sensible and latent heat.
2   Temperature of air leaving grilles, calculated or assumed.
3   Mass of air to be circulated.
4   Temperature loss in ducts.
5   Output of heaters, washers, humidifiers, coolers.
6   Boiler or heater size.
7   Duct system and duct sizes.

1   Heating and cooling loads
Calculated with data in sections 6 and 7.

2   Supply air temperature
For heating $38°-50°C$ $(100°-120°F)$.

For cooling, inlets near occupied zones, $6°-8°C$ below room temperature $(10°-15°F)$.

For cooling, high velocity diffusing jets. $17°C$ below room temperature $(30°F)$.

3   Air quantity

$$W = \frac{H}{C(t_d - t_r)} \qquad V = \frac{H}{C\varrho(t_d - t_r)}$$

where

$W =$ mass of air (kg/s)
$V =$ volume of air (m^3/s)
$H =$ sensible heat loss or gain (kW)
$C =$ specific heat capacity of air ($= 1.01$) (kJ/kg K)
$\varrho =$ density of air ($= 1.21$) (kg/m^3)
$t_d =$ discharge temperature of air at grilles (°C)
$t_r =$ room temperature (°C)

when moisture content is limiting factor

$$W = \frac{M}{w_2 - w_1}$$

where

$W =$ mass of air (kg/s)
$M =$ moisture to be absorbed (g/s)
$w_1 =$ humidity of supply air (g/kg)
$w_2 =$ humidity of room air (g/kg)

Alternatively, the air quantity is determined by the ventilation requirements of the occupants or process in the various rooms.

It is a disadvantage of the *Combined System* that the air quantity necessary to satisfy the heating or cooling requirement is not always the same as that necessary to satisfy the ventilation requirement and an acceptable compromise is not always easy to find.

4   Temperature drop in ducts

$$WC(t_1 - t_2) = Ak\left(\frac{t_1 + t_2}{2} - t_r\right)$$

where

$W =$ mass of air flowing (kg/s)
$C =$ specific heat capacity of air ($= 1.01$) (kJ/kg K)
$A =$ area of duct walls (m^2)
$k =$ heat loss coefficient of duct walls (kW/m^2 K)
$t_1 =$ initial temperature in duct (°C)
$t_2 =$ final temperature in duct (°C)
$t_r =$ surrounding room temperature (°C)
$k = 5.68 \times 10^{-3}$ kW/m^2 K for sheet metal ducts
$\ \ = 2.3 \times 10^{-3}$ kW/m^2 K for insulated ducts.

For large temperature drops the logarithmic mean temperature should be used. The equation then becomes

$$WC(t_2 - t_1) = Ak\frac{(t_1 - t_r) - (t_2 - t_r)}{\log_e(t_1 - t_r)/(t_2 - t_r)}$$

5   Heaters, washers, humidifiers, coolers
Units with required combination of air quantity, heating or cooling capacity, humidifying or dehumidifying capacity to be selected from manufacturers' catalogues.

6   Boiler

$$B = H(1 + X)$$

where

$B =$ boiler rating (kW)
$H =$ total heat load of all heater units in system (kW)
$X =$ margin for heating up and design uncertainties (0.15 to 0.20)

Boiler with correct rating to be selected from manufacturers' catalogues.

7 Duct sizes

$$v = \frac{Q}{A}$$

$$p_t = p_1 + p_2 + p_3$$

$$p_1 = il$$

$$i = \frac{2f\varrho v^2}{d}$$

$$p_2 = \sum \frac{Kv^2\varrho}{2}$$

where

$v =$ air velocity (m/s)
$Q =$ air volume (m^3/s)
$A =$ cross section of duct (m^2)
$p_t =$ total pressure loss in system (N/m^2)
$p_1 =$ pressure loss in ducts due to friction (N/m^2)
$p_2 =$ pressure loss in fittings (N/m^2)
$p_3 =$ pressure loss in apparatus (filters, heaters, etc.) (N/m^2)
$i =$ duct friction resistance per unit length (N/m^2 per m run)
$f =$ friction factor, which is a function of Reynolds number
$K =$ coefficient of resistance for fitting
$\varrho =$ density of air (kg/m^3)
$d =$ diameter of duct (m)

i can be obtained from Chart 4.
For rectangular ducts the equivalent diameter must be used

$$d = 1.26 \sqrt[5]{\frac{(ab)^3}{a+b}}$$

where

$d =$ equivalent diameter (m)
$a, \ b =$ sides of rectangular duct (m)

For standard air

$$\varrho = 1.21 \, \text{kg/m}^3$$

$$p_2 = K\left(\frac{v}{1.29}\right)^2 \text{N/m}^2 \text{ with } v \text{ in m/s}$$

Ventilation rates, occupancy known

Type of building	Fresh air supply m^3/s per person	Type of building	Fresh air supply m^3/s per person
Assembly halls	0.014	Schools	0.014
Factories	0.02–0.03	Shops	0.02
Hospitals, general	0.025	Theatres	0.014
contagious		Areas where heavy	
diseases	0.05	smoking can occur	0.028
Offices	0.016		

Ventilation rates, occupancy unknown

Type of building	Air changes per hour	Type of building	Air changes per hour
Assembly halls	5–10	Laundries	10–15
Baths	5–8	Libraries	3–4
Boiler rooms	4	Offices	3–8
Cinemas	5–10	Museums	3–4
Conference rooms	6–10	Restaurants	7–15
Department stores	3–8	Sports halls	6
Dry stores	10	Supermarkets	3–8
Engine rooms	4	Swimming pools	5–10
Factories	6	Theatres	5–10
Garages	6	Toilets	6–10
Kitchens	10–60		
Laboratories	4–15		

Garage ventilation
Two thirds total extract at high level, one third at low level

Bathroom and W.C. ventilation
Six air changes per hour or 0.018 m^3/s per room.

To provide a standby service two fans with an automatic changeover switch are installed.

Proprietary units incorporating two fans with automatic changeover are widely used. Alternatively individual fans can be joined by ducting and the changeover control supplied separately. Typical schemes for this are

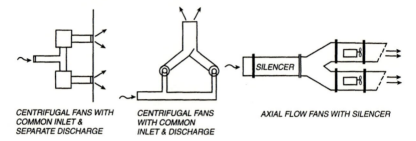

CENTRIFUGAL FANS WITH
COMMON INLET &
SEPARATE DISCHARGE

CENTRIFUGAL FANS
WITH COMMON
INLET & DISCHARGE

AXIAL FLOW FANS WITH SILENCER

Theoretical velocity of air (due to natural draught)

$$V = 4.43\sqrt{\frac{h(t_c - t_o)}{273 + t_o}}$$

$$V = 8.02\sqrt{\frac{h(t_c - t_o)}{460 + t_o}}$$

$V =$ theoretical velocity (m/s)
$h =$ height of flue (m)
$t_c =$ temperature of warm air column (°C)
$t_o =$ temperature of outside air (°C)
$V =$ in ft/s
$h =$ in ft
$t_c =$ in °F
$t_o =$ in °F

Air velocities and equivalent pressures

$$p = \frac{V^2 \varrho}{2}$$
$$= 0.6V^2$$

$p =$ velocity pressure N/m^2
$V =$ velocity m/s
$\varrho =$ density of air $= 1.2$ kg/m^3

$$h = \frac{V^2 \varrho}{2g} \frac{1}{18\,720}$$
$$= \frac{V^2}{16\,000\,000}$$

$h =$ velocity head in water gauge
$V =$ velocity ft/min
$\varrho =$ density of air $= 0.075$ lb/ft^3

Filters

Dust load for filters

	mg/m^3
Rural and suburban districts	0.45–1.00
Metropolitan districts	1.0–1.8
Industrial districts	1.8–3.5

Types of filter

(a) *Washers*

Overall length	about 20 m
Air velocity through washer	2.5 m/s
Water quantity required	0.5 to 0.8 litre per m^3 air
Water pressure required for spray nozzles	140–170 kN/m^2
Water pressure required for flooding nozzles	35–70 kN/m^2

(b) *Dry filters*
Felt, cloth, cellulose, glass, silk, etc. without adhesive liquid
  (i)   Panel type – disposable
      Air velocity   0.1–1.0 m/s
      Resistance   25–250 N/m^2
  (ii) Continuous roll – self cleaning
      Air velocity   2.5 m/s
      Resistance   30–175 N/m^2

(c) *Viscous filters*
  (i)   Panel type – cloth with viscous fluid coating – washable or disposable
      Plates about   500 mm×500 mm
      Air velocity   1.5–2.5 m/s
      Resistance   20–150 N/m^2
  (ii) Continuous roll - continuously moving, self cleaning
      Air velocity   2.5 m/s
      Resistance   30–175 N/m^2

(e) *Electrostatic precipitators*
      Cleaned automatically
      Air velocity   1.5–2.5 m/s
      Resistance   negligible

(f) *Absolute*
      Dry panel with special coating – disposable or self cleaning
      Air velocity   2.5 m/s
      Resistance   250–625 N/m^2

Resistance of ducts. (Allowance for surface conditions)

Surface	Chart reading to be multiplied by
Asbestos cement	0.8
Asphalted cast iron	6.0
Aluminium	0.8
Brickwork	4
Concrete	2
Fibreglass	0.8
PVC	0.8
Sheet iron	1.5
Sheet steel	1.0

Coefficients of resistance (for ductwork fittings)

FITTING	K	FITTING	K
SHARP 90° BEND	1.3	EXIT TO ROOM	1.0
90° BEND WITH VANES	0.7	ENTRY FROM ROOM	0.5
ROUNDED 90° BEND r/w < 1	0.5	ABRUPT REDUCTION	$0.5 - (V_1/V_2)^2$ APPLIED TO V.H. OF V_2
ROUNDED 90° BEND r/w > 1	0.25	ABRUBT ENLARGEMENT	$(1 - V_2/V_1)^2$ APPLIED TO V.H. OF V_1
SHARP 45° BEND	0.5	TAPERED ENLARGEMENT α ≤ 8°	$1.5 \times (1 - V_2/V_1)^2$ APPLIED TO V.H. OF V_1
ROUNDED 45° BEND r/w < 1	·0.2	TAPERED ENLARGEMENT α > 8°	$(1 - V_2/V_1)^2$ APPLIED TO V.H. OF V_1
ROUNDED 45° BEND r/w > 1	0.05	TAPERED REDUCTION	0
FLOW TO BRANCH	0.3 APPLIED TO V.H. IN THE BRANCH	GRILLES	RATIO OF FREE AREA TO TOTAL SURFACE
			0.7 — 3
			·0.6 — 4
			0.5 — 6
			0.4 — 10
			0.3 — 20
			0.2 — 50

Pressure drop in apparatus. (Usually given by manufacturers)

Apparatus	Average pressure drop	
	(N/m^2)	(in w.g.)
Filters	50 to 100	$\frac{3}{16}$ to $\frac{3}{8}$
Air washers	50 to 100	$\frac{3}{16}$ to $\frac{3}{8}$
Heater batteries	30 to 100	$\frac{1}{8}$ to $\frac{3}{8}$

Recommended velocities for ventilating systems

	Velocity			
	Public buildings		Industrial plant	
Service	m/s	ft/min	m/s	ft/min
Air intake from outside	2.5–4.5	500–900	5–6	1000–1200
Heater connection to fan	3.5–4.5	700–900	5–7	1000–1400
Main supply ducts	5.0–8.0	1000–1500	6–12	1200–2400
Branch supply ducts	2.5–3.0	500–600	4.5–9	900–1800
Supply registers and grilles	1.2–2.3	250–450	1.5–2.5	350–500
Low level supply registers	0.8–1.2	150–250	–	–
Main extract ducts	4.5–8.0	900–1500	6–12	1200–2400
Branch extract ducts	2.5–3.0	500–600	4.5–9	900–1800

Velocities in natural draught extract systems should be 1–3 m/s (200–600 ft/min).

Thickness of ducts

Rectangular

Longest Side		Thickness	
		Up to 1000 pa	*1001 pa to 2000 pa*
mm	*in*	*mm*	*mm*
400	15	0.6	0.8
600	24	0.8	0.8
800	32	0.8	0.8
1000	40	0.8	0.8
1250	48	1.0	1.0
1600	63	1.0	1.0
2000	78	1.0	1.2
2500	96	1.0	1.2
3000	118	1.2	–

Circular diameter		Thickness		
		Spirally wound	*Straight seamed*	
		up to 2000 pa	*up to 1000 pa*	*1001 pa to 2000 pa*
mm	*in*	*mm*	*mm*	*mm*
80	3	0.4	0.6	0.8
160	6	0.5	0.6	0.8
200	8	0.5	0.6	0.8
315	12	0.6	0.6	0.8
500	20	0.6	0.8	0.8
800	32	0.8	0.8	1.0
1000	40	1.0	1.0	1.2
1500	60	1.2	1.2	1.2

Ducts outside buildings exposed to atmosphere should be 0.2 mm thicker.

Beaufort wind scale

Beaufort No.	Description of wind	Observation	Wind speed		
			mph	ft/min	m/s
0	Calm	Smoke rises vertically	0–0.3	0–25	0–0.15
1	Light air	Direction of wind shown by smoke drift but not by wind vanes	0.3–6	25–525	0.15–2.7
2	Light breeze	Wind felt on face; leaves rustle; ordinary vanes moved by wind	6–8	525–700	2.7–3.6
3	Gentle breeze	Leaves and small twigs in constant motion; wind extends light flag	8–16	700–1400	3.6–7.2
4	Moderate breeze	Raises dust and loose paper; small branches moved	16–20	1400–1800	7.2–8.9
5	Fresh breeze	Small trees in leaf begin to sway	20–28	1800–2500	8.9–12.5
6	Strong breeze	Large branches in motion; whistling heard in telegraph wires	28–32	2500–2800	12.5–14.5
7	Moderate gale	Whole trees in motion; inconvenience felt when walking into wind	32–44	2800–3900	14.5–20
8	Gale	Twigs broken off trees; generally impedes progress	44–50	3900–4400	20–22
9	Strong gale	Slight structural damage, e.g. slates and chimney pots removed from roofs	50–62	4400–5450	22–28
10	Storm	Trees uprooted; considerable structural damage	62–70	5450–6150	28–31
11	Violet storm	Widespread damage	70–82	6150–7200	31–37
12	Hurricane		>82	>7200	>37

Chill effect is the cooling effect of air movement. It is defined as the reduction in dry bulb air temperature which would give the same cooling effect in still air.

Air velocity		Chill effect	
m/s	ft/min	°C	°F
0.1	20	0	0
0.25	50	0.5	2
1.5	300	4	7
3	600	6	10.5
5	1000	7	13
8	1575	8	15
10	2000	9	16

Circular equivalents of rectangular ducts for equal friction

$$d = 1.26 \sqrt[5]{\frac{(ab)^3}{a+b}}$$

Duct Sides	100	150	200	250	300	350	400	450	500	600	700	800	900	1000	1100	1200	1400	1500	1600	1800	2000	2200	2400	2600	2800	3000
100	110																									
150	134	165																								
200	153	190	219																							
250	170	211	245	274																						
300	185	230	268	300	329																					
350	198	247	288	324	355	384																				
400	210	263	307	345	379	410	439																			
450	221	277	324	365	401	435	465	494																		
500	231	290	340	383	422	457	490	520	548																	
600	250	315	369	417	460	499	535	569	600	658																
700	267	337	396	447	494	537	576	613	647	710	768															
800	283	357	420	475	525	571	613	653	690	758	820	878														
900	297	375	442	501	554	603	648	690	730	803	869	930	987													
1000	311	393	463	525	581	632	680	724	767	844	915	980	1040	1100												
1100	323	409	482	547	606	660	710	757	801	883	958	1030	1090	1150	1210											
1200	335	424	500	568	629	686	738	788	834	920	998	1070	1140	1200	1260	1320										
1400	357	453	534	607	673	734	791	844	895	988	1070	1150	1230	1300	1360	1420	1540									
1500	367	466	550	625	694	757	816	871	923	1020	1110	1190	1270	1340	1410	1470	1590	1650								
1600	377	479	565	643	713	779	839	896	950	1050	1140	1230	1310	1380	1450	1520	1640	1700	1760							
1800	396	503	594	676	751	819	884	944	1000	1110	1210	1300	1380	1460	1530	1610	1740	1800	1860	1970						
2000	414	525	621	707	785	857	925	989	1050	1160	1260	1360	1450	1530	1610	1640	1830	1900	1960	2080	2190					
2200	430	546	646	736	818	893	964	1030	1090	1210	1320	1420	1510	1600	1690	1770	1920	1990	2050	2180	2300	2410				
2400	446	566	670	763	848	927	1000	1070	1140	1260	1370	1480	1580	1670	1760	1840	2000	2070	2140	2280	2400	2520	2630			
2600	460	585	693	789	877	959	1040	1110	1180	1300	1420	1530	1630	1730	1820	1910	2070	2150	2220	2360	2500	2620	2740	2850		
2800	474	603	714	814	905	990	1070	1140	1210	1350	1470	1580	1690	1790	1880	1980	2150	2230	2300	2450	2590	2720	2840	2960	3070	
3000	488	620	735	838	932	1020	1100	1180	1250	1390	1510	1630	1740	1850	1950	2040	2220	2300	2380	2530	2680	2810	2940	3060	3180	3290

Duct sizes below thick line have aspect ratios greater than 4:1 and should be avoided for reasons of friction and noise.

Natural ventilation

Relies on natural forces of wind and temperature differences to generate flow of air.

Advantages

    Absence of mechanical components.

    No plant room needed.

    Reduction in building energy requirements.

Disadvantages

    Close control not practicable.

    Incoming air cannot be filtered.

    Difficult to exclude external noise.

    Paths for flow of air must form part of architectural building design.

    Cost saving of mechanical plant may be offset by increased cost of special building components.

Typical schemes

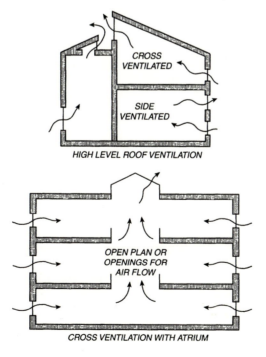

HIGH LEVEL ROOF VENTILATION

CROSS VENTILATION WITH ATRIUM

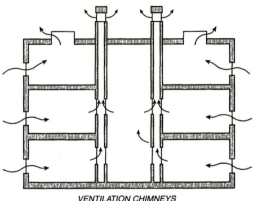

VENTILATION CHIMNEYS

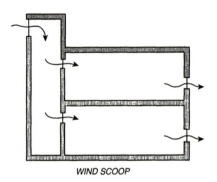

WIND SCOOP

Mixed mode ventilation

Mechanical ventilation is used in winter to avoid cold draughts and natural ventilation in summer to save energy.

Design requirements

Inlets and air paths to be arranged to avoid draughts.

Internal heat gains to be kept to a minimum.

Solar heat gain through glazing to be limited.

A higher room temperature is acceptable than with full air conditioning.

Daylighting and natural ventilation are improved by increased floor to ceiling height.

Building shape to ensure adequate wind pressure difference between inlet and outlet opening.

Air leakage through building joints and materials to be kept to a minimum. For controllable nature ventilation air tightness not to exceed 5 m³/hr per m² facade.

Openings to be adjustable and well sealed when closed to allow for difference in wind and stack effects between winter and summer.

Fire dampers to be provided at openings in fire walls.

Consideration to be given to sound insulation.

Supply air to interior areas not to pick up contamination from occupants and equipment in external areas.

Design procedure

1 Air flow requirements for summer cooling.
2 Selection of inlet and outlet ventilation openings.
3 Driving pressure due to wind and stack effects.
4 Resistance of flow path.
5 Size of ventilation openings to give required flow rate.

1 Air flow

Flow required for cooling calculated as for mechanical ventilation.

Computer programs take account of effect of thermal mass of building over 24 hour period.

For natural ventilation room air temperature can be higher than with full air conditioning.

2 Openings

Windows
Trickle ventilators
Louvres
Underfloor ducts
Chimney type stacks
Dampers - manual or automatic control

3 Driving pressure

(a) Wind pressure

Driving pressure due to wind is difference between wind pressures on inlet and outlet sides of building.

$$P_w = 0.5 C_p \varrho V_w{}^2$$

where

P_w = Wind pressure (N/m²)
C_p = Wind coefficient on wall
V_w = Wind speed (m/s)
ϱ = Density of air (kg/m³)

Data on wind coefficients has been published by Air Infiltration and Ventilation Centre, Coventry.

(b) Stack effect
Driving pressure due to stack effect

$$\Delta P_s = \varrho_i g h \frac{(T_i - T_o)}{T_o}$$

where

ΔP_s = stack driving pressure (N/m^2)
ϱ_i = density of internal air (kg/m^3)
g = acceleration of gravity = 9.81 m/s^2
h = difference in height of inlet and outlet openings (m)
T_i = internal temperature ($^\circ$K)
T_o = outside temperature ($^\circ$K)

4 Resistance of flow path
To be calculated as for large ducts with low air velocity.

5 Size of openings
Pressure drop across openings = total driving pressure-pressure drop through building.

$$A = \frac{Q}{C_d} \left[\frac{\varrho}{2\Delta P} \right]^{1/2}$$

where
A = area of opening (m^2)
Q = air flow (m^3/s)
C_d = discharge coefficient (can be taken as 0.61)
ϱ = density of inside air (kg/m^3)
ΔP = pressure drop across opening (N/m^2)

Fume and dust removal

Equipment for industrial exhaust systems

A Suction hoods, booths, or canopies for fume and dust collection, or suction nozzles, or feed hoppers for pneumatic conveying.

B Conveying, ducting or tubing.

C Fan or exhauster to create the necessary pressure or vacuum for pneumatic conveying.

D Dust separator, for separating the conveyed material from the conveying air.

Classification of schemes

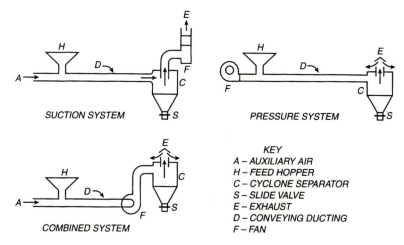

SUCTION SYSTEM

PRESSURE SYSTEM

COMBINED SYSTEM

KEY
A – AUXILIARY AIR
H – FEED HOPPER
C – CYCLONE SEPARATOR
S – SLIDE VALVE
E – EXHAUST
D – CONVEYING DUCTING
F – FAN

Pneumatic conveying plants are suitable for conveyance of material in powdered form or in solids up to 50 mm size, dry: not more than 20% moisture, not sticking.

Efficiency of pneumatic conveying plants is low but compensated by easy handling, free of dust.

Suction type – Distance of conveying up to 300 m difference in heights up to 40 m. Required vacuum 200 to 400 mm mercury.

Pressure type – Distance of conveying above 300 m working pressure up to 40 kN/m^2. Advantage: possibility of conveying material over long distance by connecting more systems in series.

Working pressure above 40 kN/m^2 not suitable, because of high running cost.

Types of hoods

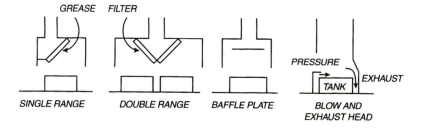

SINGLE RANGE

DOUBLE RANGE

BAFFLE PLATE

BLOW AND
EXHAUST HEAD

A flow of air into the hood resists cross draughts which would carry fumes and convected heat from appliances into the room.

Projection of hood beyond range 300 mm. Projection to be beyond open position of oven doors. Underside of hood 2000–2100 mm above floor level. Extract air velocity across face of hood 0.25 to 1.5 m/s. Mechanical supply air 85% of extract, 15% by natural infiltration.

Face velocity of mesh type grease filter 2–5 m/s, for baffle type 5 m/s.

Recommended velocities through top hoods and booth, subject to cross draughts in m/s

Canopy hood, open	1.0–1.5	Canopy hood, double	5.0
Canopy hood, closed 1 side	0.9–1.0	Booths, through 1 side	0.5–0.75
Canopy hood, closed 2 sides	0.75–0.9	Laboratory hoods,	
Canopy hood, closed 3 sides	0.5–0.75	through doors	0.25–0.35

Coefficients of entry and velocity. Pressure loss of duct extraction hoods

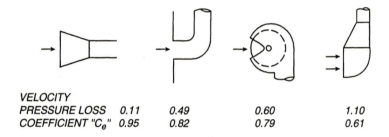

VELOCITY				
PRESSURE LOSS	0.11	0.49	0.60	1.10
COEFFICIENT "C_e"	0.95	0.82	0.79	0.61

Flow of air into a hood

$Q = 1.3\ C_e A_t \sqrt{h_t}$
Q = Air volume m³/s
C_e = Entrance coefficient
A_t = Area of throat, m²
h_t = Static suction in throat, N/m²

$Q = 4000\ C_e A_t \sqrt{h_t}$
Q = Air volume ft³/min
C_e = Entrance coefficient
A_t = Area of throat, ft²
h_t = Static suction in throat, inches w.g.

Coefficient of entry

$$C_e = \sqrt{\frac{h_v}{h_t}}\ (h_v = \text{velocity pressure})$$

Entrance loss into hood

$$C_e h_t = \frac{(I - C_e^2)}{C_e^2} h_v$$

The transporting velocity for material varies with the size, specific gravity and shape of the material

(Dalla Valle)

Vertical lifting velocity

$$V = 10.7 \frac{s}{s+1} \times d^{0.57} \qquad V = 13\,300 \frac{s}{s+1} \times d^{0.57}$$

Horizontal transport velocity

$$V = 8.4 \frac{s}{s+1} \times d^{0.40} \qquad\qquad V = 6000 \frac{s}{s+1} \times d^{0.40}$$

V = Velocity m/s
s = Specific gravity of material
d = Average dia of largest particle in mm

V = Velocity ft/min
s = Specific gravity of material
d = Average dia of largest particle in in.

Friction loss of mixture

$$\frac{F_m}{F_a} = 1 + 0.32 \left(\frac{W_s}{W_a} \right)$$

where

F_m = Friction loss of mixture
F_a = Friction loss of air
W_s = Mass of solid
W_a = Mass of air

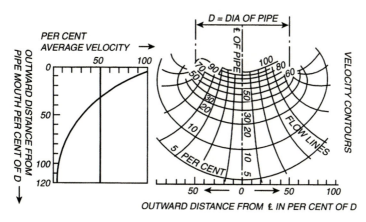

Velocity contours and flow directional lines in radial plane of circular suction pipe.

CHART 4. DUCT SIZING FOR VENTILATION SCHEMES

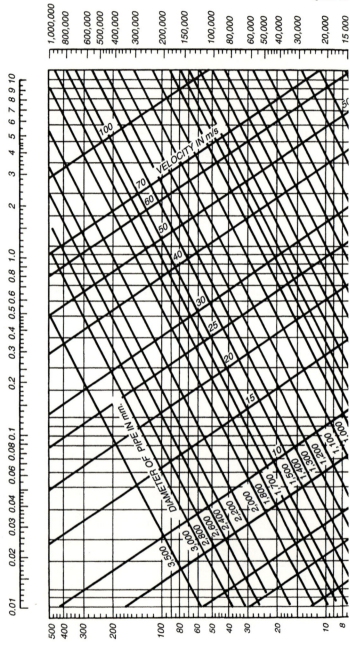

→ FRICTION IN INCHES OF WATER PER 100 ft.

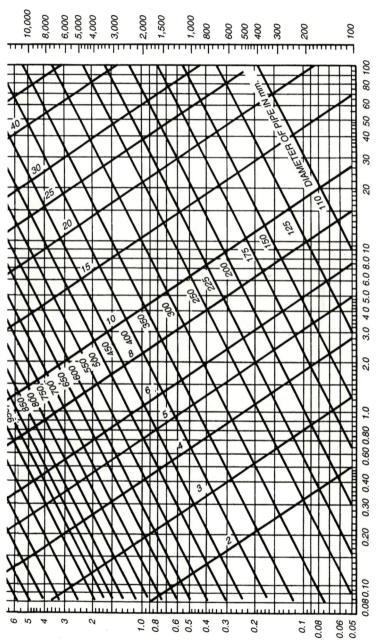

AIR VOLL ◄ ────────

10,000
8,000
6,000
5,000
4,000
3,000
2,000
1,500
1,000
800
600
500
400
300
200
100

DIAMETER OF PIPE IN mm.

─────► FRICTION IN N/m² PER m RUN R M.

0.080.10   0.20   0.30 0.40 0.60 0.80 1.0   2.0   3.0 4 0 5.0 6.0 8.0 10   20   30   40 50 60   80 100

AIR VOLL ◄ ────────

CHART 4a. DUCT SIZING FOR HIGH AIR VELOCITIES

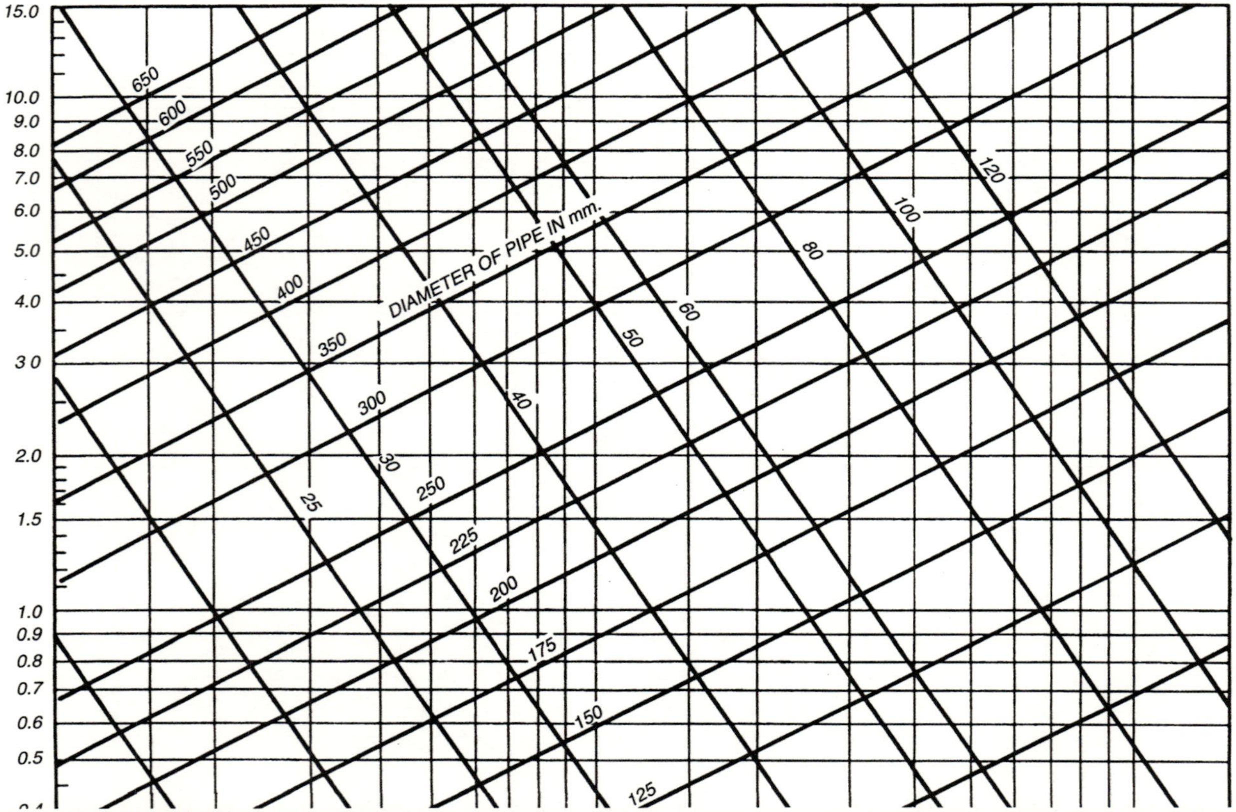

DIAMETER OF PIPE IN mm.

120
100
80
60
50
40
30
25

650
600
550
500
450
400
350
300
250
225
200
175
150
125

15.0
10.0
9.0
8.0
7.0
6.0
5.0
4.0
3.0
2.0
1.5
1.0
0.9
0.8
0.7
0.6
0.5

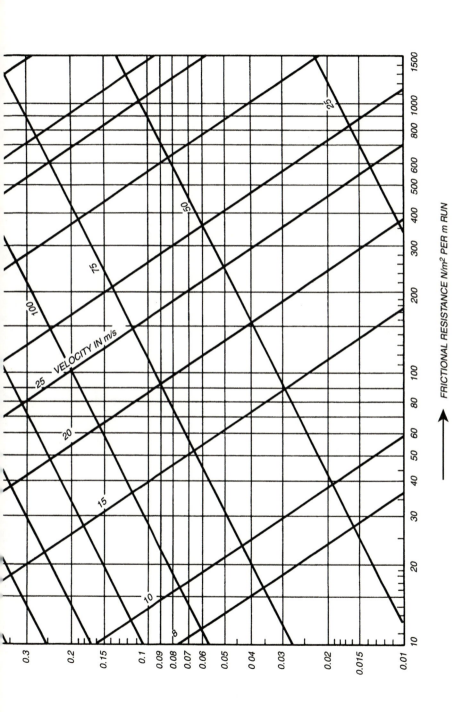

VELOCITY IN m/s

FRICTIONAL RESISTANCE N/m² PER m RUN

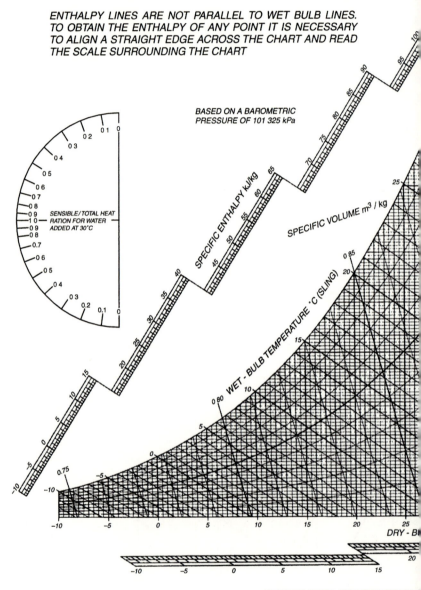

CHART 5. PSYCHROMETRIC CHART FOR AIR

ENTHALPY LINES ARE NOT PARALLEL TO WET BULB LINES.
TO OBTAIN THE ENTHALPY OF ANY POINT IT IS NECESSARY
TO ALIGN A STRAIGHT EDGE ACROSS THE CHART AND READ
THE SCALE SURROUNDING THE CHART

BASED ON A BAROMETRIC
PRESSURE OF 101 325 kPa

SENSIBLE/TOTAL HEAT
RATION FOR WATER
ADDED AT 30°C

SPECIFIC ENTHALPY kJ/kg

SPECIFIC VOLUME m³ / kg

WET - BULB TEMPERATURE °C (SLING)

DRY - B

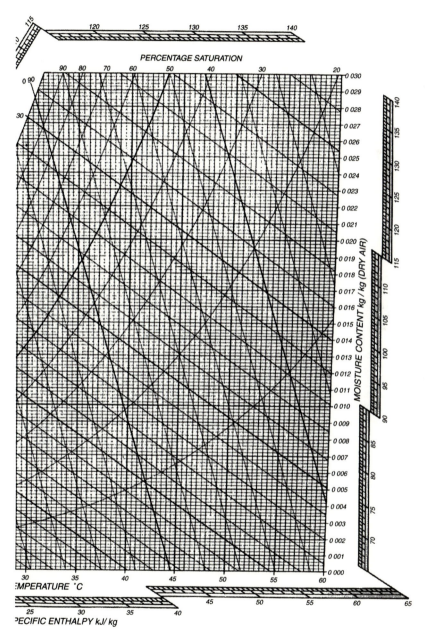

PERCENTAGE SATURATION

MOISTURE CONTENT kg / kg (DRY AIR)

TEMPERATURE °C

SPECIFIC ENTHALPY kJ/ kg

THE CHARTERED INSTITUTION OF BUILDING SERVICES ENGINEERS

Carrying velocities. (For dust extraction and pneumatic conveying)

Material	m/s	ft/min
Ashes, powdered clinker	30–43	6500–8500
Cement	30–46	6500–9000
Coal, powdered	20–28	4000–5500
Coffee beans	15–20	3000–4000
Cork	17–28	3500–5500
Corn, wheat, Rye	25–36	5500–7000
Cotton	22–30	4500–6500
Flour	17–30	3500–6500
Grain dust	10–15	2000–3000
Grinding and foundry dust	17–23	3500–4500
Jute	22–30	4500–6500
Lead dust	20–30	4500–6500
Leather dust	8–12	1800–2500
Lime	25–36	5500–7500
Limestone dust	10–15	2000–3500
Metal dust	15–18	3500–4000
Oats	22–30	4500–6500
Plastic moulding powder	15–17	3000–3500
Plastic dust	10–12	2000–2500
Pulp chips	22–36	4500–7000
Rags	22–33	4500–6500
Rubber dust	10–15	2000–3000
Sand	30–46	6000–9000
Sandblast	17–23	3500–4500
Sawdust and shavings, light	10–15	2000–3000
Sawdust and shavings, heavy	17–23	3500–4500
Textile dust	10–15	2000–3500
Wood chips	20–25	4500–5500
Wool	22–30	4500–6000

Minimum particle size for which various separator types are suitable

Gravity	200 microns (1 micron = 0.001 mm)
Inertial	50 to 150
Centrifugal, large dia cyclone	40 to 60
Centrifugal, small dia cyclone	20 to 30
Fan type	15 to 30
Filter	0.5
Scrubber	0.5 to 2.0
Electrical	0.001 to 1.0

Size of particles

Outdoor dust	0.5 microns
Sand blasting	1.4
Foundry dust	1.0 to 200
Granite cutting	1.4
Coal mining	1.0
Raindrops	500 to 5000
Mist	40 to 500
Fog	1 to 40
Fly ash	3 to 70
Pulverised coal	10 to 400

Drying

Weight of air to be circulated

$$W = \frac{X}{w_2 - w_1}$$

W = Mass of air to be circulated (kg/s)
X = Mass of water to be evaporated (kg/s)
w_1 = Absolute humidity of entering air (kg/kg)
w_2 = Absolute humidity of leaving air (kg/kg)

The relative humidity of the air leaving the dryer is usually kept below 75%.

Heat amount

Total heat amount = 1 Heat for evaporating moisture
2 Heat for heating of stock
3 Heat-loss due to air change
4 Heat transmission loss of drying chamber

Water content of various materials

Material	Original per cent	Final per cent	Material	Original per cent	Final per cent
Bituminous coal	40–60	8–12	Hides	45	0
Earth	45–50	0	Glue	80–90	0
Earth, sandy	20–25	0	Glue, air dried	15	0
Grain	17–23	10–12	Macaroni	35	0
Rubber goods	30–50	0	Soap	27–35	25–26
Green hardwood	50		Starch	38–45	12–14
Green softwood	30–50		Starch, air dried	16–20	12–14
Air dried hardwood	17–20	10–15	Peat	85–90	30–35
Air dried softwood	10–15		Yarn, washing	40–50	0
Cork	40–45	10–15			

Drying temperatures and time for various materials

Material	Temperature °C	Temperature °F	Time hr	Material	Temperature °C	Temperature °F	Time hr
Bedding	66–88	150–190		Hides, thin	32	90	2–4
Cereals	43–66	110–150		Ink, printing	21–150	70–300	
Coconut	63–68	145–155	4–6	Knitted fabrics	60–82	140–180	
Coffee	71–82	160–180	24	Leather,			
Cores, oil sand	150	300	0.5	thick sole	32	90	4–6
Films, photo	32	90		Lumber:			
Fruits, vegetable	60	140	2–6	Green,			
Furs	43	110		hardwood	38–82	100–180	3–180
Glue	21–32	70–90	2–4	Green,			
Glue size on				softwood	71–105	160–220	24–350
furniture	54	130	4	Macaroni	32–43	90–110	
Gut	66	150		Matches	60–82	140–180	
Gypsum wall				Milk	120–150	250–300	
board				Paper glued	54–150	130–300	
Start wet	175	350		Paper treated	60–93	140–200	
Finish	88	190		Rubber	27–32	80–80	6–12
Gypsum blocks	175–88	350–180	8–16	Soap	52	125	12
Hair goods	66–88	150–190		Sugar	66–93	150–200	0.3–0.!
Hats, felt	60–82	140–180		Tannin	120–150	250–300	
Hops	49–82	120–187		Terra cotta	66–93	150–200	12–96

Defogging plants

The defogging of rooms is carried out by blowing in dry, hot air and exhausting humidified air.

Mass of water evaporated from open vats

$$\frac{W}{A} = (0.037 + 0.032\,\nu) \times 10^{-3}(p_s - p_w)$$

where

W = mass of water evaporated, kg/s
A = surface area, m^2
ν = velocity of air over surface, m/s
p_s = pressure of saturated vapour at temperature of water, kPa
p_w = actual pressure of water vapour in the air, kPa

Mass of air to be circulated

$$G = \frac{W}{(w_2 - w_1)}$$

where

G = Mass of air, kg/s
W = Mass of water vapour to be removed, kg/s
w_1 = Original absolute humidity of air, kg/kg
w_2 = Final absolute humidity of air, kg/kg

Amount of heat

$$H = Gc\,(t_i - t_o)$$

where

H = Amount of heat, without fabric loss of room or other losses, W
G = Mass of air (see above), kg/s
t_i = Inside air temperature, °C
t_o = Outside air temperature, °C
c = Specific heat capacity of air = 1.012×10^3 J/Kg °C

12 Air conditioning

Design procedure for air conditioning

1 Cooling load calculation
   (a) Sensible heat load due to
      (i) heat gain through walls, etc.
      (ii) solar radiation.
      (iii) heat emission of occupants.
      (iv) infiltration of outside air.
      (v) heat emission of lights and machinery.
   (b) Latent heat load due to
      (i) moisture given off by occupants.
      (ii) infiltration of outside air.
      (iii) moisture from process machinery.
2 Selection of air treatment process. For processes and psychrometric chart see pages 80, 81 and Chart 5.
3 Determination of air quantities.
4 Layout and sizing of ducts.
5 Determination of capacities of air treating units, allowing for heat gains in ducts.
6 Determination of refrigerator and boiler duties.
7 Determination of pump and fan duties.

Methods of cooling air

1 Spray type washer.
2 Surface type cooler
   (i) Indirect. By heat exchange with water which has been cooled by a refrigerant.
   (ii) Direct. By heat exchange in evaporator of a refrigerator system.

Methods of refrigeration

1 Compression system
   Hot compressed vapour leaves a compressor and is liquified in a condenser by heat exchange with cooling water or air. The liquid refrigerant then passes through an expansion valve and the low pressure liquid enters the evaporator. It absorbs heat from the medium to be cooled and is vaporised. The vapour enters the compressor and is raised to a higher pressure.

2 Absorption system
   A solution of water in a solvent is raised to a high pressure and heated which causes the dissolved water to vaporise. The vapour is liquified in a condenser and then passes through an expansion valve. Now at low pressure the water enters the evaporator, absorbs heat from the medium to be cooled and vaporises. The vapour returns to be absorbed in the solvent.

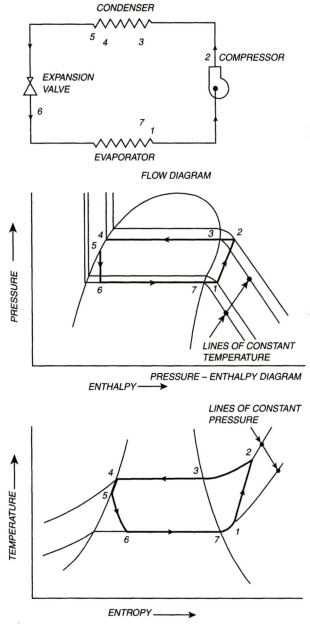

CONDENSER

5 4 3

2 COMPRESSOR

EXPANSION
VALVE

6

7 1

EVAPORATOR

FLOW DIAGRAM

PRESSURE

4

5

6 7 1

LINES OF CONSTANT
TEMPERATURE

ENTHALPY

PRESSURE – ENTHALPY DIAGRAM

LINES OF CONSTANT
PRESSURE

TEMPERATURE

2

4 3

5

6 7 1

ENTROPY

TEMPERATURE – ENTROPY DIAGRAM

VAPOUR COMPRESSION CYCLE

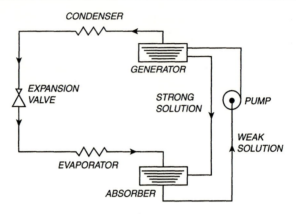

BASIC ABSORPTION CYCLE

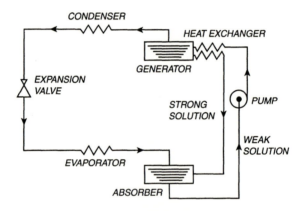

ABSORPTION CYCLE WITH SOLUTION HEAT EXCHANGER

Types of system

1　Cooling only (comfort cooling).
2　Cooling or heating.
3　Cooling or heating with control of humidity (full air conditioning).

In all systems heat is removed from the conditioned space and rejected to atmosphere outside the building.

Types of compressor

1. RECIPROCATING   Pistons driven in cylinders by crankshaft. Suction and discharge valves are thin plates which open and close easily. Most widely used type. Step control by unloading cylinders.

2. CENTRIFUGAL   Similar in construction to centrifugal pump. Vaned impeller rotates inside a casing and gas pressure is increased by centrifugal action. Suitable for very large capacities. Infinitely variable control by:

   a)　Variable speed drive
   b)　Suction throttle valve
   c)　Variable inlet guide vanes

3. SCREW   Two meshing helically shaped screws rotate and compress gas as the volume between them decreases towards the discharge side. Reliable, efficient and comparatively cheap. Used for larger duties. Infinitely variable control by bypassing partly compressed gas back to suction inlet.

4. ROTARY   Rotor with blades sliding radially is eccentric to casing. As it turns gas is swept into a smaller volume. Has few parts and can be relatively quiet and vibration free. Used for small duties such as wall or window units.

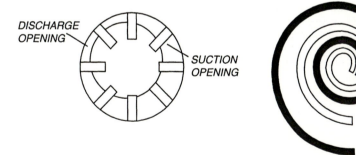

ROTARY COMPRESSOR   SCROLL COMPRESSOR

5. SCROLL   Two helically shaped scrolls are interleaved in such a way that as they rotate the space between them decreases from the suction to the discharge openings and thus compresses the gas. Suitable for medium to large commercial applications.

Free cooling

At low ambient air temperatures chilled water to air conditioning systems can be cooled directly by cold water from the cooling tower and the refrigeration plant can be switched off.

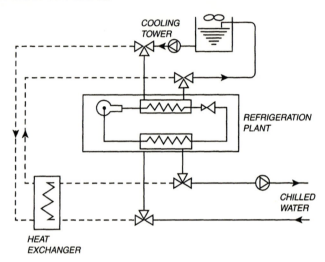

DOTTED LINES SHOW CIRCUIT AT LOW AMBIENT TEMPERATURES
THREE WAY VALVES ARE CONTROLLED BY AMBIENT AIR WET BULB TEMPERATURE

Units

Cooling is expressed in the same units as heating, namely kW or Btu/hr.

Another unit much used formerly was the Ton of Refrigeration. This was the cooling produced when one American ton of ice melted at 32°F in 24 hours. Since the latent heat of melting ice at 32°F is 144 Btu/lb

$$
\begin{aligned}
1 \text{ ton of refrigeration} &= 2000 \text{ lb} \times 144 \text{ Btu/lb in 24 hours} \\
&= 288\ 000 \text{ Btu in 24 hours} \\
&= 12\ 000 \text{ Btu/hr} \\
&= 3.517 \text{ kW}
\end{aligned}
$$

Air washer

Air washers are sheet metal, or sometimes brick or concrete chambers, in which air is drawn through a mist caused by spray nozzles and then through eliminators to remove particles of water not evaporated into the air. The water for the spray nozzles is recirculated by a pump and can be heated or cooled. A tempering heater is installed before, and a reheating battery after the air washer.

General data

Cleaning efficiency   70% on fine dust
                              98% on coarse dirt

Air velocity through washer   2-3 m/s 450-550 ft/min
Resistance   50-140 N/m^2 0.2-0.5 in water gauge
Water pressure for sprays   100-170 kN/m^2 15-25 lb/in^2
Water quantity   0.45-0.55 l/m^3 air 3-3.5 gal per 1000 ft^3 air

Humidifying efficiency

$$E = \frac{t_1 - t_2}{t_1 - t_w} \times 100\%$$

where

$t_1 = $ initial dry bulb temperature
$t_2 = $ final dry bulb temperature
$t_w = $ initial wet bulb temperature.

Typical efficiencies obtained are

60-70% with one bank of nozzles downstream
65-75% with one bank of nozzles upstream
85-100% with two banks of nozzles.

Shell and tube cooler

Shell and tube coolers consist of plain or finned tubes in an outer shell. Air flows through the shell and a liquid coolant (water, brine or refrigerant) flows through the tubes. The air can be dehumidified as well as cooled by being cooled below its dew point so that part of the moisture is condensed.

Surface area of cooler

$$A = \frac{H}{U(t_a - t_m)}$$

where

$A = $ area of cooling surface (m^2)
$H = $ cooling rate (kW)
$U = $ heat transfer coefficient (kW/m^2 K)
$t_a - t_m = $ log mean temperature difference between air and coolant (K)

Direct dehumidification

Classification
1 Adsorption type.
2 Absorption type.

Adsorption type

In adsorption systems the humidity is reduced by adsorption of moisture by an adsorbent material such as silica gel or activated alumina. Adsorption is a physical process in which moisture is condensed and held on the surface of the material without any change in the physical or chemical structure of the material. The adsorbent material can be reactivated by being heated, the water being driven off and evaporated.

The adsorption system is particularly suitable for dehumidification at room temperature and where gas or high pressure steam or hot water is available for reactivation.

Temperature for reactivation   160–175°C 325–350°F
Heat required for reactivation   4800–5800 kJ/kg water removed
                                              2100–2500 Btu/lb water removed

Silica gel SiO_2, is a hard, adsorbent, crystalline substance; size of a pea; very porous.
Voids are about 50% by volume.
Adsorbs water up to 40% of its own mass
Bulk density 480–720 kg/m^3
Specific heat capacity 1.13 kJ/kg K

Activated alumina is about 90% aluminium oxide, Al_2O_3; very porous
Voids about 50–70% by volume
Adsorbs water up to 60% of its own mass
Bulk density 800–870 kg/m^3
Specific heat capacity 1.0 kJ/kg K

Absorption type

In absorption systems the humidity is reduced by absorption of moisture by an absorbent material such as calcium chloride solution. Absorption involves a change in the physical or chemical structure of the material and it is not generally practicable to reactivate the material.

Humidification
Classification
1   Sprayed coil.
2   Spinning disc.
3   Steam humidifier.

Sprayed coil
Water is discharged into the air stream flowing onto a finned cooling coil. The discharge is through banks of high-pressure nozzles which produce a finely divided spray of water droplets. The fins of the cooling coil provide additional surface on which the water evaporates into the air. Eliminator plates at the exit from the humidifier prevent carry over of unevaporated water droplets. Excess water is collected at the base of the unit and recirculated to the nozzles by a pump. Air velocity through unit 2.0–3.5 m/s.

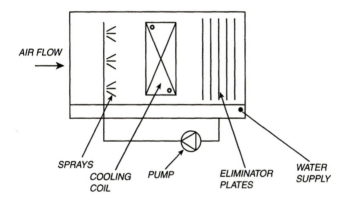

Spinning disc
Water flows as a fine film over the surface of a rapidly revolving vertical disc. It is thrown off the disc by centrifugal action onto a toothed ring and broken up into fine particles.

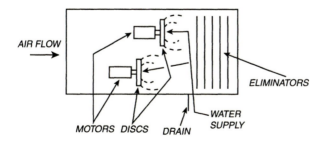

Steam humidifier

Electrode-type steam generator delivers low pressure steam to a perforated distributor pipe in the air stream. Because the water is boiled this method reduces the risk of transferring water-borne bacteria to the air.

Precautions against legionellosis

Humidifiers creating a spray should be supplied with treated and disinfected water which is not allowed to stand in equipment or tanks. Cooling towers to be positioned away from air inlets and populated areas, to have high efficiency drift eliminators and have the section above the pond enclosed to reduce wind pick up. Water to be treated to reduce scale, sediments and bacteria.

Duct work to be designed so that water cannot accumulate in air stream. Drains to have air break.

Air conditioning systems

1   Self-contained wall or window unit

Unit mounted in wall or window with evaporator inside room and condenser outside room.

Advantages
Low cost.
Flexible.
Simple.

Disadvantages
Short life.
Noise.
Poor control.
Poor filtration and air distribution.
Lack of fresh air supply.
Unsightly.

Applications
Small buildings, individual rooms.

2 Split direct expansion (DX) unit

Air cooled condenser is separate and remote from indoor unit. Compressor can be in either part but is usually in the outdoor unit.

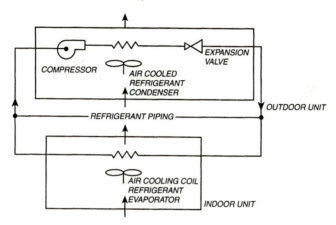

Advantages

Indoor unit need not be on outside wall.
Indoor unit can be ceiling mounted.
Silencers can be incorporated for indoor unit.
Multiple refrigerant circuits give improved control.
Relatively simple.

Disadvantages

Restriction on length of refrigerant piping between indoor and outdoor units.
Restriction on difference in level between indoor and outdoor units.
Limited fresh air supply.

Applications

Small shops, computer rooms, individual rooms or areas.

3 Split system reversible heat pump

Split system direct expansion with changeover valves enabling functions of condensing and evaporating coils to be reversed.

Advantages

Similar to split cooling only system.
Provides winter heating as well as summer cooling.

Disadvantages

Similar to split cooling only system.
Heating and cooling capacities not independent of each other.

Applications
Small shops, individual rooms or areas.

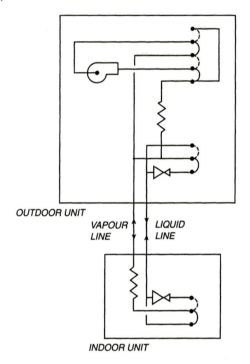

VALVES IN POSITION SHOWN IN FULL LINES
– INDOOR UNIT EVAPORATES } ROOM IS
– OUTDOOR UNIT CONDENSES } COOLED

VALVES IN POSITION SHOWN DOTTED
– INDOOR UNIT CONDENSES } ROOM IS
– OUTDOOR UNIT EVAPORATES } HEATED

For small duties the units are simplified by the use of a capillary which can operate in either direction instead of expansion valves.

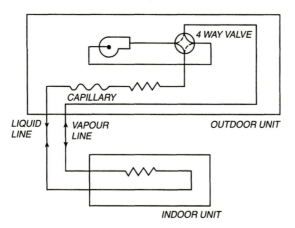

4 Water cooled unit

Self-contained indoor unit consisting of evaporator, compressor and water cooled condenser with separate outdoor cooling tower.

Advantages
> Quieter than air cooled unit.
> Flexibility in location of outdoor and indoor units.
> Better control than air cooled units.

Disadvantages
> Cooling water treatment advisable.
> Maintenance of cooling water circuit.

Applications
> Computer rooms.

5 Glycol cooled unit

Self-contained indoor unit consisting of evaporator, compressor and glycol cooled condenser with remote forced draught glycol/air heat exchanger.

Advantages
> No water treatment problems.
> Protection against freezing.

Disadvantages
> Need to ensure glycol is retained in system.

Applications
> Computer rooms.

6 Fan coil units

Chilled water or hot water is circulated from central plant to individual units in which room air is cooled or heated.

(a) *Two-pipe system.* One pair of pipes used for chilled water in summer and for hot water in winter. Suitable for continental climate with sharp difference between summer and winter. Not suitable for temperate climate as in the United Kingdom.

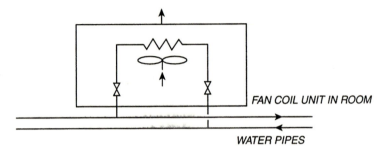

FAN COIL UNIT IN ROOM

WATER PIPES

(b) *Four-pipe system.* Separate pairs of pipes for chilled water and hot water. More expensive but more flexible control for use in temperate climates. Some rooms can be cooled while others are heated.

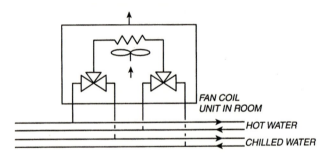

FAN COIL
UNIT IN ROOM

HOT WATER

CHILLED WATER

Advantages
Flexible.
Straightforward design.
Good control.

Disadvantages
Additional provision needed for fresh air supply.
Condense drain from each unit and/or separate provision for dehumidification.

Applications
Offices, hotel bedrooms, luxury housing, schools.

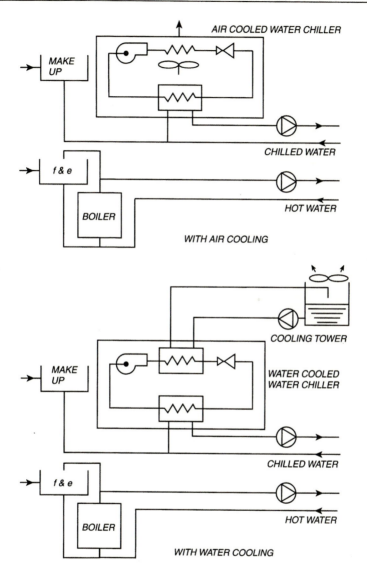

AIR COOLED WATER CHILLER

MAKE UP

CHILLED WATER

f & e

HOT WATER

BOILER

WITH AIR COOLING

COOLING TOWER

WATER COOLED WATER CHILLER

MAKE UP

CHILLED WATER

f & e

HOT WATER

BOILER

WITH WATER COOLING

CENTRAL PLANT FOR FAN COIL SYSTEM

Design parameters

Chilled water flow to fan coils	5°C–10°C
Chilled water temperature rise in fan coils	5 K–6 K
Hot water flow to fan coils	80°C
Hot water temperature drop in fan coils	10 K

7   Heat recovery units (Versatemp system from Temperature Ltd.)

Self-contained refrigeration/heat pump room units reject heat to water circulating throughout building when cooling or take heat from the water when heating. Heat rejected by units acting as coolers is supplied to units acting as heaters. Central plant to provide cooling and heating is needed to balance the cooling/heating loads.

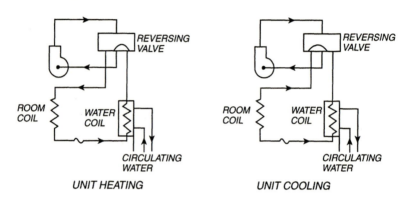

UNIT HEATING   UNIT COOLING

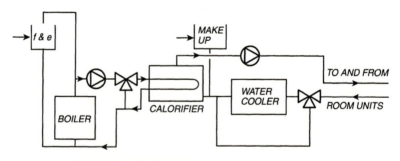

CENTRAL PLANT FOR HEAT RECOVERY SYSTEM

Advantages

Energy conservation, particularly in temperate climates.

Disadvantages

Units are larger than fan coil units.

Applications

Offices.

Design parameters

Water flow to units controlled at 27°C.
Return from individual unit

when heating 19°C
when cooling 38°C.

To achieve 27°C in summer conditions the circulating water must be cooled in a cooling tower.

Temperature Ltd offer an extended range of room units which operate with a water flow temperature of 37°C. This allows the circulating water to be cooled in a dry air blast cooler.

For water flow to units at 37°C return from individual unit

when heating 32°C
when cooling 44°C.

Disadvantage of operating at higher temperature is that room units are bigger for same duty.

8 Induction system

A central air plant delivers conditioned air through high-velocity ducting to induction units in the rooms. Water from a central plant is also supplied to the induction units. The conditioned, or primary, air supplied to the units induces room, or secondary, air through the unit. This induced secondary air passes over the water coil and is thus heated or cooled.

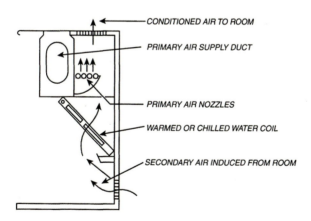

(a)   *Two-pipe changeover system*. One pair of pipes used for chilled water in summer and for hot water in winter. Not suitable for temperate climate.

(b)   *Two-pipe non-changeover system*. One pair of pipes for chilled water only, with heating by primary air only.

(c)   *Four-pipe system*. Separate pairs of pipes for chilled water and hot water. Lower running cost and better control than two-pipe non-changeover system.

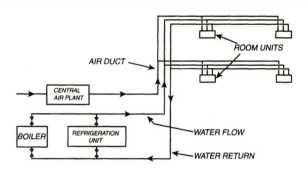

Advantages

Space saving through use of high velocity and small diameter ducts.
Low running costs.
Individual room control.
Very suitable for modular building layouts.
Central air plant need handle only part of the air treated.
Particularly applicable to perimeter zones of large buildings.
Suitable for large heat loads with small air volumes.

Disadvantages

High capital cost.
Design, installation and operation are all more complex than with fan coil
   system.
Individual units cannot be turned off.

Applications

Offices.

Design parameters

Fresh air quantity	0.012 m³/s per person or as needed for ventilation
Air velocity in primary ducts	15–20 ms
Induction unit ratio secondary air/ primary air	3 : 1
Pressure of primary air at units	200 N/m²
Hot water flow to units	80°C
temperature drop in units	10 K or as specified by manufacturer
Chilled water flow to units	5–10°C but taking into account dew point of room air
temperature rise in units	5–6 K or as specified by manufacturer

Water and air quantities and temperatures to be checked for compatibility
and required outputs at both summer and winter conditions.

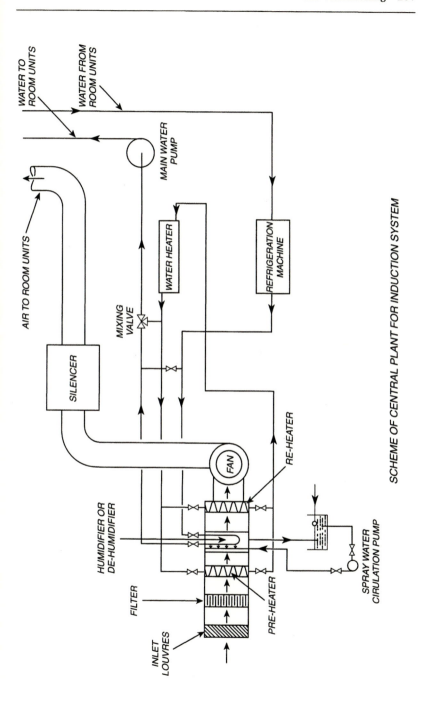

SCHEME OF CENTRAL PLANT FOR INDUCTION SYSTEM

9   All air constant volume reheat system

Central or local plant with cooler sized for latent heat cooling load and reheater to balance for sensible heat load and for winter heating. Reheater can be remote from cooler; several reheaters can be used with one cooler to give a degree of local control. Can incorporate humidifier with preheater to give complete control of discharge air temperature and humidity.

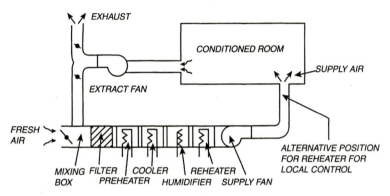

Advantages

Simple.
Free cooling available at low outdoor temperatures.
Several reheat zones can be used to improve control.
Good air distribution possible because diffusers handle constant volume.
Independent control of temperature and humidity.

Disadvantages

Wastes energy by reheat.
Expensive in both capital and running cost.
Space occupied by air ducts.
Large volume of air to be treated in central plant.
Recirculating system necessary.

Applications

Industrial, small commercial, internal areas of large buildings, houses, apartments, shopping malls, supermarkets, large stores, restaurants, theatres, cinemas, concert halls, museums, libraries, swimming pools, sports centres, clean rooms, operating theatres, large computer installations.

Design parameters

Fresh air quantity:	0.012 m³/s per person or as needed for ventilation
Air velocity:	as for ventilation systems, see Chapter 11
Supply air temperature for heating:	38°C-50°C
for cooling:	6-8 K below room temperature
Recirculating air quantity:	as required to carry heat load at specified temperature difference between room and supply air

10 Dual duct system

A central plant delivers two streams of air through two sets of ducting to mixing boxes in the various rooms. The two streams are at different temperatures.

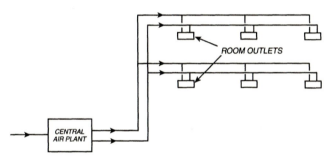

Advantages

Cooling and heating available simultaneously.
Free cooling available at low outdoor temperatures.
Individual room control — zoning not necessary.
Flexible in operation.

Disadvantages

Two sets of supply air ducting are needed, using more space.
More air has to be treated in central plant.
Recirculation system necessary.
Expensive in both capital and running costs.

Applications

Hospitals, public rooms of hotels.

11 Multizone units

Similar to dual duct system but mixing of air streams takes place at central plant for several building zones.

Advantages
Only one supply duct needed to each zone.
Free cooling available at low outdoor temperatures.

Disadvantages
Suitable only for limited number of zones.
Poor control if duties of zones differ greatly.
Recirculating system necessary.

Applications
Small buildings, groups of rooms in public buildings, swimming pools, leisure centres, libraries.

12 High-velocity air systems
Similar to all air systems but operate with high air velocities in supply ducts. Outlet boxes incorporate sound attenuators. Recirculation is usually at low velocity.

(a) *Single duct*

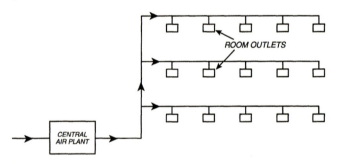

Advantages
Space saving through use of high velocity small diameter ducts.
Simple.
Zone control can be used.

Disadvantages
Large volume of air to be treated in central plant.
Individual room control not possible.
Recirculating system necessary — usually at low velocity.
Outlet attenuator boxes needed to overcome noise generated by high velocity ducting.
Higher fan pressure and fan power; increased running costs.

(b) *Dual duct*. Similar to low-velocity dual duct but with sound attenuation incorporated in outlet boxes.

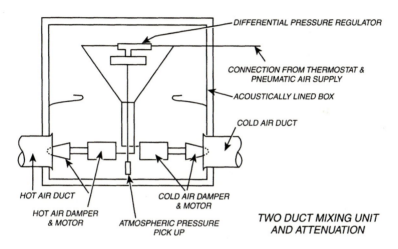

DIFFERENTIAL PRESSURE REGULATOR

CONNECTION FROM THERMOSTAT &
PNEUMATIC AIR SUPPLY

ACOUSTICALLY LINED BOX

COLD AIR DUCT

HOT AIR DUCT

HOT AIR DAMPER
& MOTOR

COLD AIR DAMPER
& MOTOR

ATMOSPHERIC PRESSURE
PICK UP

*TWO DUCT MIXING UNIT
AND ATTENUATION*

Advantages
Space saving through use of high-velocity small diameter ducts.
Individual room control – zoning not necessary.
Flexible in operation.
Can handle larger air volumes than single duct.

Disadvantages
Two sets of supply air ducting are needed, using more space.
More air has to be treated in central plant.
Recirculating system necessary – usually at low velocity.
Outlet boxes must include attenuators to overcome noise generated in high-velocity ducting.
Higher fan pressure and fan power increase running costs.

Applications
Offices, public rooms of hotels, internal areas of large buildings.

Design parameters for single and dual duct high-velocity systems
Air velocities in ducts: 15–20 m/s
Pressure at inlet to furthest unit: 100–250 N/m^2
Typical pressure at fan: 1250–1500 N/m^2
Air quantities and temperatures: as for low-velocity systems.

13 Variable air volume system
An all air system in which local control is obtained by varying volume discharged at each diffuser or group of diffusers in response to the dictates of a local thermostat. Capacity of supply and extract fans is reduced as total

system volume requirement falls at part load. Fans controlled by:

(a) Variable speed.
(b) Variable blade pitch.
(c) Variable inlet guide vanes.
(d) Disc throttle on fan outlet.

Satisfactory operation is critically dependent on the design and performance of the terminal diffuser units. Manufacturer's data must be adhered to.

Advantages

Efficient part load operation.
Individual room or area control.
Unoccupied areas can be closed off with dampers.

Disadvantages

Special provision needed for heating.
Extra controls needed to maintain minimum fresh air supply to terminals operating at low load.
Complexity of controls.
Cannot provide full control of humidity.

Methods of providing heating

(a) *Perimeter heating with VAV cooling only to core of building*
Simple.
Running cost uneconomic.
Controls may cause perimeter heating to add unnecessarily to cooling load.

(b) *Dual-duct system*
Expensive in capital cost.
Complicated and difficult to control.
Two sets of supply air ducting, using more space.

(c) *Reheater in each terminal unit*
Simple and effective.
Reheating cooled air reduces the economic operation which is chief attraction of variable air volume.

Applications

Offices, hospitals, libraries, large stores, schools.

Design parameters

Air velocities in ducts	10–15ms
Supply air temperature for cooling	9–11 K below room temperature
for heating	max 35°C
Throw and spacing of units	in accordance with manufacturer's recommendations
Turn down ratio	as advised by manufacturer 30%–20% can be achieved.

14   Displacement ventilation

Cooled air is introduced at low level at low outlet velocity. It spreads across the room at floor level and is drawn in to feed plumes of warmed air rising from occupants and equipment heat sources. It is extracted at high level. Low level inlets may be on walls or columns or grilles in a false floor.

Advantages
Removal of contaminants at source by rising plumes gives better room air quality.
Higher supply air temperature requires less refrigeration.
Simple plant and ductwork layout.

Disadvantages
Separate provision needed for heating, usually perimeter heating.
Possibility of draughts at ankle level near outlets.
Repositioning of outlets if partitioning or furniture layout is changed.

Applications
Industrial, commercial, offices, theatres, cinemas.

Design parameters
Supply temperature: 2–3 K below room temperature
Discharge velocity: 0.1–0.3 m/s
Outlets to be selected in accordance with manufacturer's data.

15   Chilled ceiling

Cool water is circulated through panels in the ceiling or through beams which may be exposed or recessed. Panels in the ceiling cool occupants by radiation from occupants to cool surface. Chilled beams have a radiant effect but also cool rising warm air and produce a convective downflow of cool air. This enables beams to have a greater cooling effect than ceiling panels.

Advantages
Cooled rather than chilled water requires less refrigeration.
Ventilation needed only for fresh air supply, therefore smaller volume.
Takes up no floor space.
Cooling by radiation permits higher room air temperature.
Low maintenance.

Disadvantages
Risk of condensation at cold surface requires control of room humidity.
Insulation needed on top of ceiling panels and beams.
Other provision needed for heating, usually perimeter heating.

Applications
Offices, public buildings.

Design parameters

Water flow temperature:	14–15°C
Water temperature rise:	2–3 K
Cooling effect:	30–80 W/m² floor area
Temperature difference, room to ceiling surface:	4–8 K
Temperature difference, water to ceiling surface:	2–3 K

Actual data to be agreed with ceiling or beam manufacturer according to application.

16 Variable refrigerant volume

Similar to split direct expansion system but several indoor units are connected by a common system of refrigerant piping to one outdoor unit. Local control is obtained by varying the flow of refrigerant at each indoor unit. Compressor output is reduced as total system requirement falls at part load. A heat recovery version is possible in which hot refrigerant from units which are cooling is passed to units which are heating.

Design in accordance with manufacturer's data

Advantages

Efficient part load operation.
Individual room or area control.

Disadvantages

Separate provision may be needed for heating.
Restriction imposed by design of refrigerant piping.
Limited fresh air supply.

Applications

Offices

Ice storage

Ice is made when electric power for refrigeration is available at a low off-peak rate. Stored ice is used to chill water for air conditioning during peak times. The store can be used for whole or part of load. A store used for part load only reduces peak demand for refrigeration and allows smaller chillers to be used, running for longer at their full load and optimum efficiency.

Direct system

Direct heat exchange between refrigerant and ice/water.
Water alone used in secondary circuit.

Freezing and melting circuits separate.

Advantage: easier to maintain low chilled water temperature.

Disadvantage: refrigerant evaporator within ice store limits distance between store and chiller.

Indirect system

Intermediate circuit between refrigerant and ice/water.

Same circuit used for both freezing and melting. Intermediate circuit must contain anti-freeze.

Advantage: no restriction on distance between ice store and chiller.

Disadvantages: changeover valves needed.

Concentration of anti-freeze must be maintained.

Ice stores

Ice builder — refrigerant evaporator within tank of water. Ice builds on evaporator coils. Store discharged by water circulated through tank.

Ice bank — glycol mixture circulated through coil below $0°C$ for freezing and above $0°C$ for melting. No circulation through tank itself.

Equipment

To be selected from manufacturers' data.

Refrigerant evaporator must operate at lower temperature than for normal air conditioning.

Capacity

$$S = \frac{p \sum h}{\eta}$$

$$R = H - \frac{S}{n_1}$$

$$\text{or} \quad \frac{S}{n_2}$$

where

S = stored energy (kWhr)

p = proportion of cooling demand over cycle to be stored (= 1 for full storage)

h = load during an hour of cycle (kWhr)

η = efficiency of store (normally about 0.94)

R = chiller capacity (kW)

H = peak cooling load (kW)

n_1 = time during which cooling is required (hr)

n_2 = charging period (hr)

Controls
Output regulated by variation of flow of chilled water through store. Detection of quantity of ice in store can be used to vary timing of cycle.

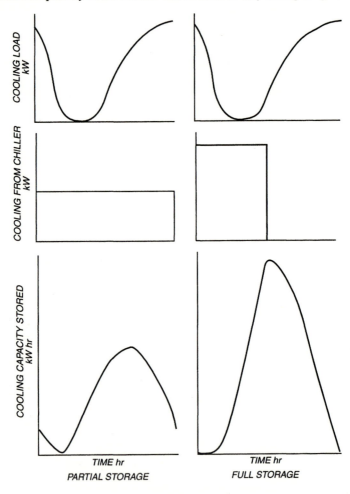

PARTIAL STORAGE

FULL STORAGE

ICE STORAGE

Properties of refrigerants

Under European legislation the use of chlorofluorocarbons is banned from 31st December 2000. The use of hydrochlorofluorocarbons is being phased out and will be banned from 1st January 2010.

The following table gives the characteristics of new and replacement refrigerants.

Refrigerant	Formula	Boiling temp. °C	Critical temp. °C	Properties	Applications
Ammonia	NH_3	−33	133	Penetrating odour, soluble in water, harmless in concentrations up to 0.33%, non-flammable, explosive, zero ozone depletion Low global warming potential	Large industrial plants
Lithium Bromide	LiBr	−	−	Soluble in alcohol and ether Soluble in water Zero ozone depletion Low global warming potential	Solvent for water in absorption systems
R134a	$CF_3 \, CH_2 \, F$	−26	101	Zero ozone depletion	Air conditioning Industrial refrigeration Domestic refrigeration Replacement for R12
R404A	CF_3CHF_2 (44%) CF_3CH_3 (52%) CF_3CH_2F (4%)	−46	72	Zero ozone depletion Non flammable Low toxicity	Cold stores and refrigerated display cabinets Replacement for R502
R407A	CH_2F_2 (20%) CHF_2CF_3 (40%) CF_3CH_2F (40%)	−42	83	Zero ozone depletion Non flammable Low toxicity	Low temperature applications Replacement for R502

Properties of refrigerants (*continued*)

Refrigerant	Formula	Boiling temp. °C	Critical temp. °C	Properties	Applications
R407C	CH_2F_2 (23%) CHF_2CF_3 (25%) CF_3CH_2F (52%)	−43	87	Zero ozone depletion Non flammable Low toxicity	Air conditioning Heat pumps Replacement for R22
R410A	CH_2F_2 (50%) CF_3CHF_2 (50%)	−52	72	Zero ozone depletion Non flammable Low toxicity Non corrosive	Air conditioning units Heat pumps Cold stores Industrial and commercial refrigeration
R507	CF_3CHF_2 (50%) CF_3CH_3 (50%)	−47	71	Zero ozone depletion Low toxicity Non corrosive	Low and medium temperature applications Refrigerated display cases Replacement for R502
CARE 40 (R290) Propane	$CH_3CH_2CH_3$	−42	97	Zero ozone depletion Low global warming potential Flammable Non toxic	Commercial and industrial refrigeration Air conditioning Heat pumps Alternative to R22 and R502
CARE 50 (R170)	$CH_3CH_2CH_3$ CH_3CH_3	−49	79	Zero ozone depletion Low global warming potential Flammable Non toxic	Commercial and process refrigeration Air conditioning Heat pumps Alternative to R22 and R502
CARE 10 (R600a) Isobutane	$CH(CH_3)_3$	−12	135	Zero ozone depletion Low global warming potential Flammable Non toxic	Small charge hermetic applications Domestic refrigeration
CARE 30	$CH(CH_3)_3$ $CH_3CH_2CH_3$	−32	106	Zero ozone depletion Low global warming potential Flammable Non toxic	Chilled food display cabinets Drinking water dispensers Alternative to R12

CARE is a trademark of Calor Gas Ltd

Former refrigerants

For reference and comparison the properties of previously common refrigerants which are now either obsolete or obsolescent are listed below.

Refrigerant	Formula	Boiling temp. °C	Critical temp. °C	Properties	Applications
R12	CCl_2F_2	-30	112	Non flammable Non corrosive Stable	Small plants with reciprocating compressors
R11	CCl_3F	9	198	Non flammable Non corrosive Stable	Commercial plants with centrifugal compressors
R22	$CHClF_2$	-41	96	Non flammable Non toxic Non corrosive Stable	Packaged air conditioning units
R500	CCl_2F_2 (74%) CH_3CHF_2 (25%)	-33		Non flammable Non corrosive Stable	Approximately 20% more refrigeration capacity than R12. Useful when machine designed for 60 Hz had to operate on 50 Hz
R502	$CHClF_2$ (50%) $CClF_2CF_3$	-46	90	Non flamable Non toxic Non corrosive	Low temperature applications

Friction loss through fittings

The following table takes into account static regain. EL = Equivalent length of pipe.

Fitting		4 in 100 mm	6 in 150 mm	8 in 200 mm	10 in 250 mm	12 in 300 mm	14 in 350 mm	16 in 400 mm	18 in 450 mm	20 in 500 mm
	EL ft	−9	−15							
	EL m	−3	−5							
	EL ft	12	21							
	EL m	4	7							
90°	EL ft	3	4	7	10	12	15	18	21	24
90°	EL m	1	1.2	2	3	4	5	6	7	8
45°	EL ft	1	2	4	5	6	8	9	10	12
45°	EL m	0.3	0.6	1.2	1.5	1.8	2.4	3	3	4
30°	EL ft	1	1	2	3	4	5	6	7	8
30°	EL m	0.3	0.3	0.6	1	1.2	1.5	1.8	2	2.4
	EL ft	−5	−9	−13	−17	−22	−26	−31	−36	−42
	EL m	−1.5	−3	−4	−5	−7	−8	−10	−11	−13
	EL ft	12	21	30	40	52	63	75	87	100
	EL m	4	6	10	12	16	19	22	25	30
	EL ft	−13	−22	−32	−42	−54	−66	−78	−91	−105
	EL m	−4	−7	−10	−13	−16	−20	−24	−28	−32
	EL ft	13	22	32	42	54	66	78	91	105
	EL m	4	7	10	13	16	20	24	28	32
$\frac{d}{D} < 04$	EL ft	−8	−10	−11	−13	−17	−20	−24	−28	−32
$\frac{d}{D} < 04$	EL m	−2.4	−3	−3.3	−4	−5	−6	−7	−9	−10
$\frac{d}{D} \le 04$	EL ft					−9	−9	−10	−10	−10
$\frac{d}{D} \le 04$	EL m			−3	−3	−3	−3	−3	−3	−3
	EL ft	13	22	32	42	54	56	78	91	105
	EL m	4	7	10	13	16	20	24	28	32
	EL ft		−21	−30	−40	−52	−63	−75	−87	−100
	EL m		−6	−10	−12	−16	−19	−22	−25	−30
	EL ft	13	22	32	42	54	66	78	91	105
	EL m	4	7	10	13	16	20	24	28	32
	EL ft	14	23	37	49	62	76	90	106	121
	EL m	4	7	11	15	19	23	27	32	

Design temperatures and humidities for industrial processes

Industry	Process		Temperature °C	Relative humidity %
Textile	Cotton	carding	24–27	50
		spinning	15–27	60–70
		weaving	20–24	70–80
	Rayon	spinning	21	85
		twisting	21	65
	Silk	spinning	24–27	65–70
		weaving	24–27	60–70
	Wool	carding	24–27	65–70
		spinning	24–27	55–60
		weaving	24–27	50–55
Tobacco	Cigar and cigarette making		21–24	55–65
	Softening		32	85
	Stemming and strigging		24–30	70
Paint	Drying oil paints		15–32	25–50
	Brush and spray painting		15–27	25–50
Paper	Binding, cutting, drying, folding, gluing		15–27	25–50
	Storage of paper		15–27	34–45
	Storage of books		18–21	38–50
Printing	Binding		21	45
	Folding		25	65
	Pressing, general		24	60–78
Photographic	Development of film		21–24	60
	Drying		24–27	50
	Printing		21	70
	Cutting		22	65
Fur	Storage		−2 to +4	25–40
	Drying		43	–

Air curtains

Heated air is blown across a door opening to prevent or reduce ingress of cold atmospheric air.

Applications
Door-less shop fronts.
Workshop entrances.
Doors of public buildings which are frequently opened.

Temperatures
Discharge Temperature: for small installation   35-50°C
                         for large installation   25-35°C
                         Suction Temperature   5-15°C

Air velocity
Flow from above   5-15 m/s
          below   2-4 m/s
          side   10-15 m/s

Air quantity:
Quantity required depends on too many variable factors for exact calculation to be possible. The quantity should be made as large as possible consistent with practicable heat requirements. Suggested values: 2000-5000 m^3/m^2 hr of door opening. In very exposed situations or other difficult cases this can be increased to 10 000 m^3/m^2 hr.

Let V_o = quantity of air entering in absence of curtain
$\quad V$ = quantity blown by curtain
For one-sided curtain $V = 0.45 \ V_o$
For two-sided curtain $V = 0.9 \ V_o$

Example: Width of door 4 m. Height of door 2 m. Speed of outdoor air 2 m/s.

$$\therefore V_o = 4 \times 2 \times 2 = 16 \ m^3/s$$
$$\therefore V = 0.45 \times 16 = 7.2 \ m^3/s$$

Discharge velocity, say 10 m/s.

$$\therefore \text{Grille area} = \frac{7.2}{10} = 0.72 \, m^2$$

Height of grille = height of door = 2 m.

$$\therefore \text{Width of grille} = \frac{0.72}{2} = 0.36 \ m$$

13 Pumps and fans

Flow in pipes

Bernoulli's Equation can be applied between points in a pipe through which fluid is flowing, with the addition of a term to allow for energy lost from the fluid in overcoming friction.

$$\frac{p_1}{\varrho g} + \frac{U_1^2}{2g} + z_1 = \frac{p_2}{\varrho g} + \frac{U_2^2}{2g} + z_2 + h_f$$

$$h_f = \frac{p_f}{\rho g}$$

where

Subscript 1 refers to values at point 1.
Subscript 2 refers to values at point 2.
p = pressure (N/m^2)
ϱ = density (kg/m^3)
g = weight per unit mass
   = acceleration due to gravity (m/s^2)
U = velocity (m/s)
z = height above arbitrary datum (m)
h_f = friction head from point 1 to point 2 (m)
p_f = pressure necessary to overcome friction between points
   1 and 2 (N/m^2)

Fluid statics

For a liquid in equilibrium

$$p + \varrho g z = \text{const.}$$

If the datum from which z is measured is taken as the free surface of the liquid

$$z = -h$$

and

$$p = \varrho g h$$

$$\frac{p}{\varrho g} = h \text{ and is termed } \textit{pressure head}$$

$$\frac{p}{\varrho g} + z = \text{const. and is termed } \textit{piezometric head}$$

where

$p = $ pressure of liquid (N/m^2)
$\varrho = $ density (kg/m^3)
$g = $ weight per unit mass
$ = $ acceleration due to gravity (m/s^2)
$z = $ height above arbitrary datum (m)
$h = $ depth below free surface (m)

Fluid motion

The total energy per unit weight of a liquid in steady flow remains constant. This is expressed in *Bernoulli's Equation*:

$$\frac{p}{\varrho} + \frac{U^2}{2} + gz = \text{const.}$$

or

$$\frac{p}{\varrho g} + \frac{U^2}{2g} + z = \text{const.}$$

$p/\varrho g$ is the pressure head per unit weight of fluid.

$U^2/2g$ is the velocity head per unit weight of fluid.

z is the gravitational head above datum per unit weight of fluid.

where

$U = $ velocity of fluid (m/s).

Other symbols as above.

Venturimeter

A venturimeter is inserted in a pipe to measure the quantity of water flowing through it.

$$Q = \frac{C_d A_2}{\sqrt{1 - (A_2/A_1)^2}} \sqrt{2\left(\frac{p_1 - p_2}{\varrho}\right)}$$

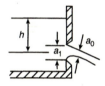

where

Q = quantity of water flowing (m³/s)
C_d = coefficient of discharge
   = 0.96 to 0.99
A = area (m²)
p = pressure (N/m²)
ϱ = density (kg/m³)

Subscripts 1 and 2 refer to values of sections 1 and 2 respectively.

Discharge of water through small orifice

$Q = v a_o$
$v = C_v \sqrt{2gh}$
$a_o = C_c a_1$
$Q = C_v C_c a_1 \sqrt{2gh}$

where

Q = quantity of water discharged (m³/s)
v = velocity at section of minimum area of jet (m/s)
a_o = area at section of minimum area of jet (m²)
a_1 = area of orifice (m²)
h = height of free surface above orifice (m)
g = weight per unit mass
   = acceleration due to gravity (m/s²)
C_v = coefficient of velocity
   $= \dfrac{\text{actual velocity}}{\text{theoretical velocity}} = 0.96 \text{ to } 0.99$
C_c = coefficient of contraction
   $= a_o/a_1 = 0.6 \text{ to } 0.7$

Velocity heads and theoretical velocities of water

$$h = \frac{v^2}{2g}$$

h = Head in m
v = Velocity in m/s
g = Gravity of earth = 9.81 m/s^2

v m/s	h m	v m/s	h m	v m/s	h m	v m/s	h m
0.01	0.0000051	0.80	0.0326	1.60	0.130	2.40	0.293
0.05	0.000127	0.85	0.0368	1.65	0.139	2.45	0.306
0.10	0.00051	0.90	0.0413	1.70	0.147	2.50	0.318
0.15	0.00115	0.95	0.046	1.75	0.156	2.55	0.331
0.20	0.00204	1.0	0.0510	1.80	0.165	2.60	0.344
0.25	0.00319	1.05	0.0561	1.85	0.174	2.65	0.358
0.30	0.00459	1.10	0.0617	1.90	0.184	2.70	0.371
0.35	0.00624	1.15	0.0674	1.95	0.194	2.75	0.385
0.40	0.00815	1.20	0.0734	2.0	0.204	2.80	0.400
0.45	0.0103	1.25	0.0797	2.05	0.214	2.85	0.414
0.50	0.0127	1.30	0.0862	2.10	0.225	2.90	0.429
0.55	0.0154	1.35	0.0930	2.15	0.236	2.95	0.444
0.60	0.0183	1.40	0.100	2.20	0.246	3.0	0.459
0.65	0.0125	1.45	0.107	2.25	0.258		
0.70	0.0250	1.50	0.115	2.30	0.269		
0.75	0.0287	1.55	0.122	2.35	0.281		

$$h = \frac{v^2}{2g}$$

h = Head in ft
v = Velocity in ft/s
g = Gravity of earth = 32.2 ft/s^2

v ft/s	h ft	v ft/s	h ft	v ft/s	h ft	v ft/s	h ft
0.1	0.0002	2.1	0.068	4.1	0.261	6.1	0.578
0.2	0.0006	2.2	0.075	4.2	0.274	6.2	0.597
0.3	0.0014	2.3	0.082	4.3	0.289	6.3	0.616
0.4	0.0025	2.4	0.089	4.4	0.301	6.4	0.636
0.5	0.0039	2.5	0.097	4.5	0.314	6.5	0.656
0.6	0.0056	2.6	0.105	4.6	0.329	6.6	0.676
0.7	0.0076	2.7	0.113	4.7	0.343	6.7	0.697
0.8	0.0099	2.8	0.122	4.8	0.358	6.8	0.718
0.9	0.0126	2.9	0.131	4.9	0.373	6.9	0.739
1.0	0.0155	3.0	0.140	5.0	0.388	7.0	0.761
1.1	0.019	3.1	0.149	5.1	0.404	7.1	0.783
1.2	0.022	3.2	0.159	5.2	0.420	7.2	0.805
1.3	0.026	3.3	0.169	5.3	0.436	7.3	0.827
1.4	0.030	3.4	0.179	5.4	0.453	7.4	0.850
1.5	0.035	3.5	0.190	5.5	0.470	7.5	0.874
1.6	0.040	3.6	0.201	5.6	0.487	7.6	0.897
1.7	0.045	3.7	0.212	5.7	0.505	7.7	0.921
1.8	0.050	3.8	0.224	5.8	0.522	7.8	0.945
1.9	0.056	3.9	0.236	5.9	0.541	7.9	0.969
2.0	0.062	4.0	0.248	6.0	0.559	8.0	0.994

Centrifugal pumps

The action of pumps is most conveniently expressed in terms of head. The rotor gives the liquid a head.

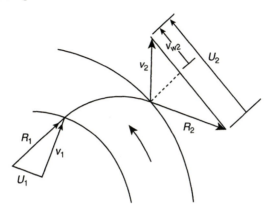

Notation

v_1 = absolute velocity of water at inlet (m/s)
v_2 = absolute velocity of water at outlet (m/s)
U_1 = tangential velocity of blade at inlet (m/s)
U_2 = tangential velocity of blade at outlet (m/s)
V_{w1} = tangential velocity of water at inlet (m/s)
V_{w2} = tangential velocity of water at outlet (m/s)
R_1 = velocity of water relative to blade at inlet (m/s)
R_2 = velocity of water relative to blade at outlet (m/s)
g = weight per unit mass = 9.81 (m/s^2)
ϱ = density (kg/m^3)
p_1 = pressure of water at inlet (N/m^2)
p_2 = pressure of water at outlet (N/m^2)
H_1 = ideal head developed by pump (m)
H_a = manometric head
   = actual head developed by pump (m)
η_m = manometric efficiency (%)
η_o = overall efficiency (%)

Normally for a pump $V_{w1} = 0$

Then

$$H_1 = \text{work done on water per unit weight} = \frac{U_2 V_{w2}}{g}$$

When the output of a pump is expressed as head of working liquid it is independent of the density of the liquid.

$$H_m = \frac{p_2 - p_1}{\varrho g} + \frac{v_2^2}{2g}$$

Actual head is less than ideal because of friction losses within pump.

$$\eta_m = \frac{H_m}{H_1} \times 100$$

Overall efficiency is lower again because of mechanical losses in bearings, etc.

$$\eta_o = \frac{H_m Q \varrho g}{S} \times 100$$

where

Q = quantity of water flowing (m^3/s)
S = power input at shaft (Nm/s)

The specific speed of a centrifugal pump is the speed at which the pump would deliver 1 m^3/s of water at a head of 1 m.

$$N_s = \frac{nQ^{1/2}}{H^{3/4}}$$

where

N_s = specific speed
n = speed (rev/min)
Q = volume delivered (m^3/s)
H = total head developed (m)

Pump laws
1   Volume delivered varies directly as speed
$$\frac{Q_1}{Q_2} = \frac{N_1}{N_2}$$
2   Head developed varies as the square of speed
$$\frac{H_1}{H_2} = \left(\frac{N_1}{N_2}\right)^2$$
3   Power absorbed varies as the cube of speed
$$\frac{S_1}{S_2} = \left(\frac{N_1}{N_2}\right)^3$$

Characteristic curves of pumps

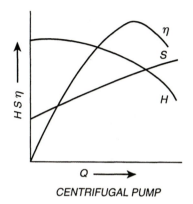

CENTRIFUGAL PUMP

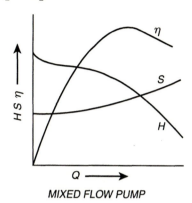

MIXED FLOW PUMP

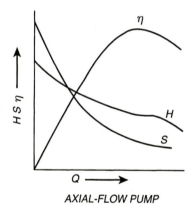

AXIAL-FLOW PUMP

Q = QUANTITY FLOWING   (m³/s)
H = HEAD DEVELOPED   (m)
S = POWER ABSORBED   (W)
η = EFFICIENCY   (%)

A centrifugal pump takes the least power when the flow is zero. It should therefore be started with the delivery valve shut.

An axial flow pump takes the least power when the flow is greatest. It should therefore be started with the delivery valve open.

Fans

1   Propeller fans and axial flow fans

Pressure for single stage up to about 300 N/m^2.
Suitable for large volumes at comparatively low pressures.

Characteristic curve for axial flow fan

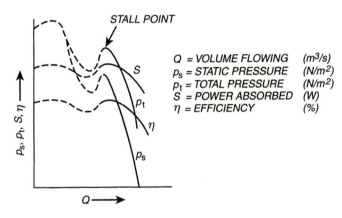

Q = VOLUME FLOWING (m^3/s)
p_s = STATIC PRESSURE (N/m^2)
p_t = TOTAL PRESSURE (N/m^2)
S = POWER ABSORBED (W)
η = EFFICIENCY $(\%)$

2   Centrifugal fans
Types of blade

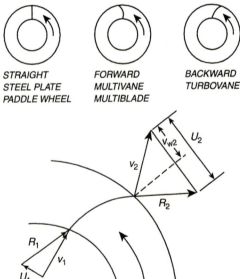

STRAIGHT
STEEL PLATE
PADDLE WHEEL

FORWARD
MULTIVANE
MULTIBLADE

BACKWARD
TURBOVANE

Notation
Suffix 1 refers to inlet.
Suffix 2 refers to outlet.

v = absolute velocity of air (m/s)
u = tangential velocity of blade (m/s)
v_w = tangential velocity of air (m/s)
R = velocity of air relative to blade (m/s)
g = weight per unit mass
  = 9.81 (m/s^2)
ϱ = density of air (kg/m^3)
p_t = total pressure (N/m^2)
p_s = static pressure (N/m^2)
p_1 = theoretical total pressure developed by fan (N/m^2)
p_a = actual total pressure developed by fan (N/m^2)
Q = volume of air (m^3/s)
S = power input to fan (W)
η = efficiency (%)

Normally

$$p_1 = U_2 v_{w2} \varrho \qquad v_{w1} = 0$$

$$p_t = p_s + \frac{v^2 \varrho}{2}$$

$$p_a = p_{t2} - p_{t1} = p_{s2} - p_{s1} + \frac{\left(v_2^2 - v_1^2\right)\varrho}{2}$$

$$\eta = \frac{P_a Q}{S} \times 100$$

Characteristic curve for centrifugal fan

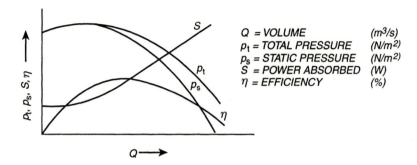

Q = VOLUME (m^3/s)
p_t = TOTAL PRESSURE (N/m^2)
p_s = STATIC PRESSURE (N/m^2)
S = POWER ABSORBED (W)
η = EFFICIENCY $(\%)$

3   Mixed flow fans

Within an axial casing the impeller hub and casing inlet form a conical passage in which the impeller blades combine axial and centrifugal actions. Downstream guide vanes turn the radial component of air velocity into axial velocity without loss of pressure. This enables a fan with an axial-type casing fitted in a straight run of ducting to develop higher pressures than a normal axial flow fan.

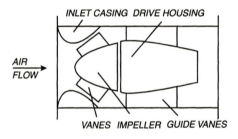

INLET CASING DRIVE HOUSING

AIR FLOW

VANES IMPELLER GUIDE VANES

CHARACTERISTIC CURVE FOR MIXED FLOW FAN

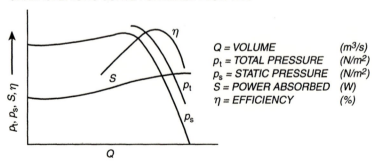

Q = VOLUME (m^3/s)
p_t = TOTAL PRESSURE (N/m^2)
p_s = STATIC PRESSURE (N/m^2)
S = POWER ABSORBED (W)
η = EFFICIENCY $(\%)$

Typical efficiencies

Small fans   0.40
Medium fans   0.60
Large fans   0.80

Fan laws

1 Volume varies directly as speed

$$\frac{Q_1}{Q_2} = \frac{N_1}{N_2}$$

2 Total pressure varies as the square of speed

$$\frac{P_{t1}}{P_{t2}} = \left(\frac{N_1}{N_2}\right)^2$$

3 Power absorbed varies as the cube of speed

$$\frac{S_1}{S_2} = \left(\frac{N_1}{N_2}\right)^3$$

Selection of fans

1 Air volume to be moved.
2 Static pressure or resistance.
3 Noise level permissible.
4 Electric supply available.

Pressures commonly used for typical systems

Public buildings, ventilation only   90-150 N/m^2
Public buildings, combined heating and ventilation   150-250 N/m^2
Public buildings, combined heating and ventilation with air
  cleaning plant   170-300 N/m^2
Factories, heating only   170-400 N/m^2
Factories, combined heating and ventilation   300-500 N/m^2

Fan discharge velocities for quiet operation

	Supply systems m/s	Extract systems m/s
Sound studios, churches, libraries	4-5	5-7
Cinemas, theatres, ballrooms	5-7.5	6-8
Restaurants, offices, hotels, shops	6-8	7-9

14 Sound

Sound. (Energy travelling as a pressure wave)

One decibel is equal to ten times the logarithm to base 10 of the ratio of two quantities.

$$I = \frac{W}{A} = \frac{p^2}{\varrho c}$$

Sound power level

$$PWL = 10 \log_{10} \frac{W}{W_o}$$

Sound intensity

$$IL = 10 \log_{10} \frac{I}{I_o}$$

Sound pressure level

$$SPL = 10 \log_{10} \frac{p^2}{p_o^2}$$
$$= 20 \log_{10} \frac{p}{p_o}$$

where

$I =$ intensity of sound (W/m^2)
$I_o =$ reference intensity (W/m^2)
$W =$ power (W)
$W_o =$ reference power (W)
$A =$ area (m^2)
$p =$ root mean square pressure (N/m^2)
$p_o =$ reference pressure (N/m^2)
$\varrho =$ density (kg/m^3)
$c =$ velocity of sound (m/s)

The usual reference levels are

$W_o = 10^{-12}$ watts
$I_o = 10^{-12}$ W/m^2
$p_o = 0.0002$ μbar $= 20 \times 10^{-6}$ N/m^2

At room temperature and at sea level $SPL = IL + 0.2$ decibels

234

Measurement of noise

Method of adding levels expressed in decibels

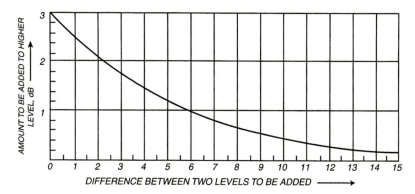

Noise rating Graphs are plotted of Sound Pressure Level (SPL) v frequency, to show how the acceptable sound level varies with frequency. What is acceptable depends on the use to which the room will be put, and so a different curve is obtained for each type of use. Each such curve is designated by an NR number.

NR No.	Application
NR 25	Concert halls, broadcasting and recording studios, churches
NR 30	Private dwellings, hospitals, theatres, cinemas, conference rooms
NR 35	Libraries, museums, court rooms, schools, hospital operating theatres and wards, flats, hotels, executive offices
NR 40	Halls, corridors, cloakrooms, restaurants, night clubs, offices, shops
NR 45	Department stores, supermarkets, canteens, general offices
NR 50	Typing pools, offices with business machines
NR 60	Light engineering works
NR 70	Foundries, heavy engineering works

NR levels (SPL. dB re 0.00002 N/m^2)

Noise rating	Octave band mid-frequency, HZ							
	62.5	125	250	500	1000	2000	4000	8000
NR10	42	32	23	15	10	7	3	2
NR20	51	39	31	24	20	17	14	13
NR30	59	48	40	34	30	27	25	23
NR35	63	52	45	39	35	32	30	28
NR40	67	57	49	44	40	37	35	33
NR45	71	61	54	48	45	42	40	38
NR50	75	65	59	53	50	47	45	43
NR55	79	70	63	58	55	52	50	49
NR60	83	74	68	63	60	57	55	54
NR65	87	78	72	68	65	62	61	59
NR70	91	83	77	73	70	68	66	64
NR75	95	87	82	78	75	73	71	69
NR80	99	91	86	82	80	78	76	74

Sound obeys the Inverse Square Law

$$p^2 = K \frac{W}{r^2}$$

where

p = root mean square pressure
K = constant
W = power
r = distance from source

or

$$SPL = PWL - 20 \log_{10} r + K'$$
$$K' = \log_{10} K = \text{constant.}$$

In air with source near ground, $K' = -8$.

For a continuing source in a room, the sound level is the sum of the direct and the reverberant sound and is given by

$$SPL = PWL + 10 \log_{10} \left[\frac{Q}{4\pi r^2} + \frac{4}{R} \right] dB$$

where

$$Q = \frac{SPL \text{ at distance } r \text{ from actual source}}{SPL \text{ at distance } r \text{ from uniform source of same power}}$$

R = Room constant = $\dfrac{S\alpha}{1-\alpha}$ m^2
S = Total surface area of room m^2
α = Absorption coefficient of walls
r in m

Coefficient of absorption α
For range of frequencies usual in ventilation applications

Plaster walls	0.01–0.03	25 mm wood wool cement	
Unpainted brickwork	0.02–0.05	on battens	0.6–0.7
Painted brickwork	0.01–0.02	50 mm slag wool or glass silk	0.8–0.9
3-plywood panel	0.01–0.02	12 mm acoustic belt	0.5–0.6
6 mm cork sheet	0.1–0.2	Hardwood	0.3
6 mm porous rubber sheet	0.1–0.2	25 mm sprayed asbestos	0.6–0.7
12 mm fibreboard on		Persons, each	2.0–5.0
battens	0.3–0.4	Acoustic tiles	0.4–0.8

Sound insulation of walls

$$\text{Transmission coefficient } \tau = \frac{\text{transmitted energy}}{\text{incident energy}}$$

Sound reduction index

$$\text{SRI} = 10 \, \log_{10}\left(\frac{1}{\tau}\right) \text{dB}$$

Empirical formula is

$$\text{SRI} = 15 \, \log(\sigma f) - 17$$

where

$\sigma = $ mass per unit area of wall (kg/m^2)
$f = $ frequency (Hz)

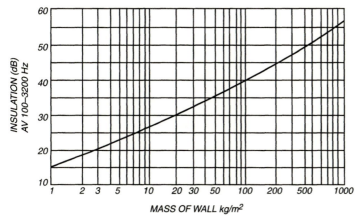

SOUND INSULATION OF SOLID WALLS ACCORDING TO MASS

Transmission through walls

$$(SPL)_1 - (SPL)_2 = SRI - 10 \log_{10}\left(\frac{S_p}{S_2\alpha_2}\right) \text{dB}$$

where

$(SPL)_1$ = sound pressure in sending room
$(SPL)_2$ = sound pressure in receiving room
SRI = sound reduction index
$S_2\alpha_2$ = equivalent absorption in receiving room
S_p = area of partition wall

Sound insulation of windows

Single/ double window	Type of window	Type of glass	Sound reduction in dB
Single	Opening type (closed)	Any glass	18–20
Single	Fixed or opening type with air-tight weather strips	24/32 oz sheet glass	23–25
		6 mm polished plate glass	27
		9 mm polished plate glass	30
Double	Opening type (closed) plus absorbent material on sides of air space	24/32 oz sheet glass 100 mm space	28
		24/32 oz sheet glass 200 mm space	31
		6 mm polished plate glass 100 mm space	30
		6 mm polished plate glass 200 mm space	33
Double	Fixed or opening type with air-tight weather strips	24/32 oz sheet glass 100 mm space	34
		24/32 oz sheet glass 200 mm space	40
		6 mm polished plate glass 100 mm space	38
		6 mm polished plate glass 200 mm space	44

Attenuation by building structure

Structure	Attenuation dB	Structure	Attenuation dB
9 in brick wall	50	Double window 50 mm spacing	30
6 in (150 mm) concrete wall	42	12 mm T & G boarded partition	26
Wood joist floor and ceiling	40	2.5 mm glass window	23
Lath and plaster partition	38		

Transmission through ducts

$$\frac{\text{Attenuation}}{\text{Duct length}} = 1.07\alpha^{1\,4}\frac{P}{A}\text{dB per ft}$$

$$= 1.07\alpha^{1\,4}\frac{P}{A}\text{dB per m}$$

where

$\alpha =$ coefficient of absorption
$P =$ perimeter of duct
$A =$ cross sectional area of duct

Approximate attenuation of round bends or square bends with turning vanes in dB

Frequency Hz	20–75	75–150	150–300	300–600	600–1200	1200–2400	2400–4800	4800–10 000
Diameter								
5 to 10 in 125 to 250 mm	0	0	0	0	1	2	3	3
11 to 20 in 251 to 500 mm	0	0	0	1	2	3	3	3
21 to 40 in 501 to 1000 mm	0	0	1	2	3	3	3	3
41 to 80 in 1001 to 2000 mm	0	1	2	3	3	3	3	3

Attenuation due to changes in area in dB

Ratio of Area S_2/S_1	Attenuation dB	Ratio of Area S_2/S_1	Attenuation dB
1	0.0	3	1.3
2	0.5	4	1.9
2.5	0.9	5	2.6

Attenuation at entry to room (end reflection loss)

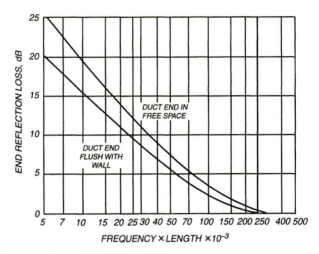

END REFLECTION LOSS
FOR RECTANGULAR OPENING LENGTH = $\sqrt{L_1 \times L_2}$
FOR CIRCULAR OPENING LENGTH = 0.9 DIA.
LENGTHS IN mm.

Sound power level (PWL) of fans

Exact data for any particular fan is to be obtained from the manufacturer. In the absence of this the following approximate expressions may be used.

$$PWL = 90 + 10 \log_{10} s + 10 \log_{10} h$$
$$PWL = 55 + 10 \log_{10} q + 20 \log_{10} h$$
$$PWL = 125 + 20 \log_{10} s - 10 \log_{10} q$$

where

s = rated motor power (hp)
h = fan static head (in water gauge)
q = volume discharged (ft^3/min)

$$PWL = 67 + 10 \log_{10} S + 10 \log_{10} p$$
$$PWL = 40 + 10 \log_{10} Q + 20 \log_{10} p$$
$$PWL = 94 + 20 \log_{10} S - 10 \log_{10} Q$$

where

S = rated motor power (kW)
p = fan static pressure (N/m^2)
Q = volume discharged (m^3/s)

Typical curves of fan frequency distribution

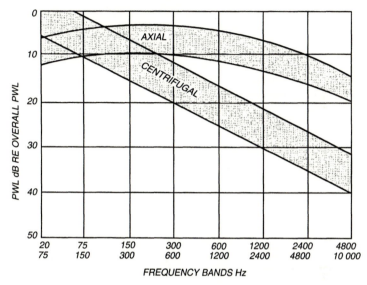

SOUND POWER LEVEL SPECTRA OF FANS

Sound absorption

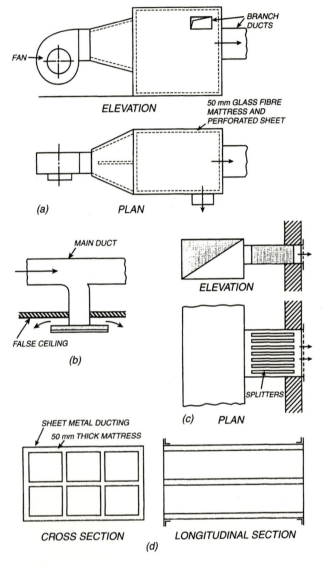

(a)   Sound absorption by increase of duct area.
(b)   Ceiling air outlet with sound-absorbing plate.
(c)   Sound absorption in branch duct with splitter.
(d)   Arrangement of splitters in main duct.

15 Labour rates

The following schedules give basic times for installation and erection of heating and ventilating equipment.

Included. Haulage of all parts into position, erection on site, surveying of builder's work, testing.

Not included. Delivery to site, travelling time, addition for overtime working.

Additions to basic time

The basic time given should be increased

For jobs under 1 week	by 40%
For jobs under 2 weeks	20%
For jobs under 3 weeks	8%
For work in existing buildings, unoccupied	5%
For work in existing buildings, occupied	15%
For work in existing building, with concealed pipes	20%

Plant	Time in man hrs
Boilers. Including all fittings	
Cast iron sectional	
up to 36 kW	20
37–75 kW	25
76–150 kW	35
151–220 kW	50
221–300 kW	60
301–450 kW	70
451–580 kW	75
581–750 kW	80
Unit construction steel boilers	
up to 110 kW	6
111–300 kW	12
301–600 kW	20

Plant	Time in man hrs
Oil burners, Pressure jet gas burners	
up to 75 kW	8
76–150 kW	12
151–300 kW	20
301–500 kW	25
Combination boiler – Calorifier sets. Including all fittings	
Boiler rating	
up to 110 kW	20
111–300 kW	25
301–600 kW	35
Calorifiers, Indirect cylinders, Direct cylinders	
Storage capacity	
up to 250 litres	9
251–550 litres	12
551–900 litres	20
901–1300 litres	30
1301–2250 litres	35
2251–3500 litres	40
3501–5000 litres	45
Electric water heaters	
up to 3 kW	4
F and E tanks, Cold water tanks	
Capacity	
up to 90 litres	6
91–225 litres	9
226–450 litres	12
451–900 litres	15
901–2000 litres	20
2001–4000 litres	30

Plant		Time in man hrs
Pumps. Complete with motor		
Direct coupled or belt driven, with supports		
Flow	Nominal size	
up to 1.5 litre/s	up to 32	10
1.6–2.5 litre/s	40–50	14
2.6–25 litre/s	65–100	18
over 25 litre/s	over 100	24
In-line pipeline pumps		
Flow	Nominal size	
up to 0.7 litre/s	up to 32	6
0.8–1.5 litre/s	40–80	10
over 1.5 litre/s	over 80	14
Flue pipes. Steel or Asbestos		
Pipes		
150 dia	(per m)	1
300 dia	(per m)	2
over 300 dia	(per m)	3
Elbows		
150 dia		0.3
300 dia		0.6
over 300 dia		1.0
Valves, Taps, Cocks		
Nominal bore		
15–32		0.5
40–50		1.0
65–100		2
over 100		3
Mixing Valves, Diverting Valves, Two-way Valves. With		
Actuators		
Nominal bore		
up to 32		8
40–50		10
65–100		12
over 100		15

Plant	Time in man hrs
Electric Starters	
All ratings	2
Three-way Cocks. For venting	
Nominal bore	
up to 50	2
65–100	3
over 100	4
Pressure Gauge. With Cock	2
Thermometer	1
Thermostat	2
Safety Valves	
Nominal bore	
up to 32	1
40–50	2
over 50	4
Radiators. Complete with 2 valves	
Heating surface	
up to 2.5 m^2	4
2.6–4.5 m^2	6
4.6–10 m^2	8
Remove and refix one radiator (for painting and decorating)	1.5
Natural Draught Convectors	
Length	
up to 1 m	6
1.1–1.5 m	7
over 1.5 m	8
Fan Convectors	
Floor standing 700 mm high	5
recessed 700 mm high	7
recessed 1800 mm high	10

Plant	Time in man hrs
Industrial Type Unit Heaters	
up to 6 kW	10
7-15 kW	15
16-30 kW	20
Gas Fired Room Heaters	
up to 2 kW	5
2.1-10 kW	7
11-18 kW	12
over 18 kW	15
Centrifugal Fans	
Complete with motor, direct coupled or belt driven, with supports	
Impeller diameter	
up to 300 mm	15
301-600 mm	20
601-1000 mm	28
1001-1200 mm	45
1201-1525 mm	50
Axial Flow Fans. Complete with casing and motor	
Diameter	
up to 300 mm	5
301-600 mm	8
601-1500 mm	10
1501-2400 mm	20
Cooling Towers. For air conditioning	
Plan area	
up to 2 m^2	15
2.1-3.5 m^2	20
3.6-6.0 m^2	45
6.1-8.0 m^2	60
8.1-15 m^2	100
15.1-20 m^2	140
20.1-25 m^2	160

Plant	Time in man hrs
Dry or Throw-away Filters	
Capacity	
up to 0.7 m³/s	1
0.71-2.0 m³/s	2
Self-cleaning Viscous Filters	
Capacity	
up to 1 m³/s	5
1.1-1.8 m³/s	7
1.9-6 m³/s	15
6.1-12 m³/s	20
12.1-22 m³/s	30
22.1-30 m³/s	40
Grease Filters	
Capacity	
up to 0.5 m³/s	1
0.6-1.0 m³/s	2
1.1-1.3 m³/s	3.5
1.4-3.0 m³/s	5
Package Air Handling Units. Consisting of filter, preheater, cooler, humidifier, reheater, fan and silencer	
Capacity	
up to 0.15 m³/s	6
0.16-0.3 m³/s	10
0.4-0.5 m³/s	15
0.6-0.8 m³/s	20
0.9-2 m³/s	30
2.1-3 m³/s	43
3.1-4.5 m³/s	70
4.6-6 m³/s	110
6.1-8 m³/s	150

Plant	*Time in man hrs*
Grilles and Registers	
Long side	
up to 100 mm	1
101–450 mm	2
over 450 mm	3
Air Dampers. In ventilation ducting	
Diameter (or equivalent diameter for rectangular dampers)	
up to 100 mm	1
101–200 mm	2
201–500 mm	3
501–1000 mm	5
1001–1700 mm	8
1701–2000 mm	10
Actuator or Motor for Motorised Dampers	
All ratings	7.5

	Time in man hr/m
Pipes. Including brackets and fittings	
Nominal bore	
up to 20 mm	0.5
25–32 mm	0.75
40–100 mm	1
150 mm	1.8
200 mm	2.5
250 mm	3
Pipe Lagging. Rigid or flexible sectional	
Nominal bore of pipe	
up to 150 mm	0.5
over 150 mm	0.75

	Time in man hr/m^2
Equipment Lagging	
Flat	0.5

	Time in man hrs/m
Steel Ducts for ventilation systems, including supports and brackets and all fittings	
Diameter (or equivalent diameter for rectangular ducts)	
up to 200 mm	3
201–300 mm	6
301–500 mm	10
501–750 mm	18
751–1000 mm	25
1001–1200 mm	40
1201–1700 mm	50
1701–2000 mm	70

	Time in man hours/tonne
Ventilation and Air Conditioning Equipment Not separately detailed, or as alternative to times given above	90

16 Bibliography

The following list is intended as a guide for readers who require more theoretical treatment of the topics on which data is presented in this book. It is a selection of books which are both useful and generally accessible, but does not claim to be exhaustive. Some of the books mentioned are out of print; they are included because they are available in libraries and contain material which is still useful.

Handbooks
CIBSE Guides. Chartered Institution of Building Services Engineers, London
Handbook of Air Conditioning Design. Carrier Air Conditioning, Biggin Hill, Kent
Kempe's Engineering Yearbook, Morgan-Grampian, London
Machinery's Handbook, The Industrial Press, New York
Mark's Standard Handbook for Mechanical Engineers, McGraw-Hill, New York
Newnes Mechanical Engineers Pocket Book, 2nd ed. (1997)
Powell, M.J.V., *House Builder's Reference Book*, Butterworth-Heinemann, Oxford (1979)

Heating, ventilating, air conditioning
Allard, F., (ed.), *Natural Ventilation in Buildings*, James & James (1998)
Awbi, H. W., *Ventilation of Buildings*, E. & F. N. Spon (1991)
Barton, J.J., *Small Bore Heating and Hot Water Supply for Small Dwellings*, 2nd ed., Butterworth-Heinemann, Oxford (1970)
Barton, J.J., *Electric Floor Warming*, Butterworth-Heinemann, Oxford (1967)
Bedford, Thomas, *Bedford's Basic Principles of Ventilation and Heating*, ed. F.A. Chrenka, 3rd ed., H.K. Lewis, London (1974)
Chadderton, D., *Air Conditioning: A Practical Introduction*, 2nd ed., E. & F. N. Spon (1997)
Chadderton, D., *Building Services Engineering*, 3rd ed., E. & F. N. Spon (2000)
Clifford, G.E., *Modern Heating & Ventilating System Design*, Prentice Hall (1993)
Cooper, W.B., Lee, R.E. and Quinlan, R.A., *Warm Air Heating for Climate Control*, 2nd ed., Prentice Hall (1991)
Croome-Gale, D.J. and Roberts, T.M., *Air Conditioning and Ventilation of Buildings*, 2nd ed. (1981)
Clements-Croome, D., *Naturally Ventilated Buildings*, E. & F. N. Spon (1997)
Faber and Kell's *Heating and Air Conditioning of Buildings*, 8th ed., Martin, P.L., and Oughton, D.R., Architectural Press (1995)
Jones, W.P., *Air Conditioning Engineering*, 4th ed., Arnold, London (1994)
Jones, W.P., *Air Conditioning Applications and Design*, 2nd ed., Arnold, London (1997)
Kreider, J.F. and Rabb, A., *Heating and Cooling of Buildings*, McGraw-Hill (1994)
Kut, D., *Heating and Hot Water Services in Buildings*, Pergamon Press, Oxford (1968)

Kut, D., *Warm Air Heating*, Pergamon Press, Oxford (1970)

Kut, D. and Hare, G., *Applied Solar Energy*, Butterworth-Heinemann, Oxford (1983)

Mackenzie-Kenney, C., *District Heating*, Pergamon Press, Oxford (1979)

McQuiston, F.C. and Parker, J.D., *Heating, Ventilating and Air Conditioning*, 4th ed., Wiley, New York

Moss, K., *Heating and Water Services Design in Buildings*, E. & F. N. Spon (1996)

Santamouris, M. and Asimakopoulos, *Passive Cooling of Buildings*, James & James (1996)

Sherratt, A.F.C., *Air Conditioning System Design for Buildings*

Wang, S.K., *Handbook of Air Conditioning and Refrigeration*, McGraw-Hill (1993)

Heat pumps

Heap, R.D., *Heat Pumps*, 2nd ed., E. & F. N. Spon, London (1983)

McMullan, T. and Morgan, R., *Heat Pump*, Hilger, Bristol (1981)

Miles, L., *Heat Pumps, Theory and Service*, Delmar Publishers (1994)

Reay, D.A. and MacMichael, D.B.A., *Heat Pumps: Design and Application* (1979)

Sherratt, A.F.C. (ed.), *Heat Pumps for Buildings*, Hutchinson, London (1984)

Von Cube, H.W. and Steimle, F., *Heat Pump Technology*, Butterworth-Heinemann, Oxford (1981)

Heat, heat transfer and thermodynamics

Dunn, P.D. and Reay, D.A., *Heat Pipes*, 3rd ed. (1982)

Eastop and McConkey, *Applied Thermodynamics for Engineering Technologies*, 5th ed., Longman (1993)

Fishenden, M. and Saunders, O.A., *An Introduction to Heat Transfer*, Oxford University Press (1950)

Joel, R., *Basic Engineering Thermodynamics*, 5th ed., Longman (1997)

Long, C.A., *Essential Heat Transfer*, Longman (1999)

Moss, K., *Heat and Mass Transfer in Building Services Design*, E. & F. N. Spon (1998)

Rogers, G.F.C. and Mayhew, Y.R., *Engineering Thermodynamics*, 4th ed., Longman, London (1992)

Sherwin, K., *Introduction to Thermodynamics*, Chapman & Hall (1993)

Refrigeration

Dossat, R.J., *Principles of Refrigeration*, 4th ed., Prentice Hall (1997)

Langley, B.C., *Refrigeration and Air Conditioning*, 2nd ed., Reston Publishing (1982)

Stoecker, W.F., *Industrial Refrigeration Handbook*, McGraw-Hill (1998)

Hydraulics

Evett, J.B. and Liu, C., *Fundamentals of Fluid Mechanics*, McGraw-Hill (1987)

Fox, R.W. and McDonald, A.T., *Introduction of Fluid Mechanics*, 5th ed., Wiley (1998)

Massey, B., revised by Ward-Smith, J., *Mechanics of Fluids*, 7th ed., Stanley Thornes (1998)

Sherwin, J. and Horsley, M., *Thermofluids*, Chapman & Hall (1996)

Combustion

Brame and King, *Fuels, Solid, Liquid and Gaseous*, Edward Arnold, London (1967)

Gilchrist, J.D., *Fuels, Furnaces and Refractories*, Pergamon Press, Oxford (1977)

Fans

Bleier, F.B., *Fan Handbook*, McGraw-Hill (1998)

Eck, I.B., *Fans*, translated from German, Pergamon Press, Oxford (1973)

Osborne, W.C., *Fans*, 2nd ed., Pergamon Press, Oxford (1977)

Wallis, R.A., *Axial Flow Fans and Ducts*, Wiley, New York (1983)

Wood's, *Practical Guide to Fan Engineering*, 3rd ed., Wood's of Colchester (1978)

Pumps

Anderson, H.H., *Centrifugal Pumps*, 3rd ed., Trade & Technical Press, Morden, Surrey (1980)

British Pump Manufacturers' Association, *Pump User's Handbook*, Trade & Technical Press, Morden, Surrey (1978)

De Kovats, A. and Desmur, G., *Pumps, Fans and Compressors*, translated from French by R.S. Eaton (1958)

Karassik, I.J., *Pump Handbook*, McGraw-Hill, New York (1986)

Lobanoff, V.S. and Ross, R.R., *Centrifugal Pumps, Design & Application*, 2nd ed., Gulf Publishing (1982)

Pumping Manual, 8th ed., Trade & Technical Press, Morden, Surrey (1988)

Warring, R.J., *Pumps: Selection, Systems and Applications*, 2nd ed., Trade & Technical Press, Morden (1984)

Sound

Croome, D.J., *Noise, Buildings & People* (1977)

Everest, F.A., *The Master Handbook of Acoustics*, 3rd ed., TAB Books (McGraw Hill) (1994)

Ghering, W.L., *Reference Data for Acoustic Noise Control* (1978)

Iqbal, M.A., *The Control of Noise in Ventilation Systems* (1977)

Porges, G., *Applied Acoustics*, Edward Arnold, London (1977)

Sharland, I., *Wood's Practical Guide to Noise Control*, Wood's of Colchester (1972)

Smith, B.J., Peters, R.J. and Owen, J., *Acoustics and Noise Control*, 2nd ed., Longman (1996)

Piping

M.W. Kellog Co., *Design of Piping Systems*, Wiley, New York (1965)

Kentish, D.N.W., *Industrial Pipework*, McGraw-Hill, London (1982)

Pearson, G.H., *Application of Valves and Fittings*, Applied Science Publishers, London (1981)
Piping Handbook, 7th ed., McGraw-Hill (2000)
Smith, P.R. and van Laan, T.J., *Piping and Pipe Support Systems*, McGraw-Hill, New York (1987)

Welding
American Welding Society, *Welding Handbook*, 7th ed., Macmillan, London (1970–78)
Cary, H.B., *Modern Welding Technology*, Prentice Hall, Englewood Cliffs (1989)
Davies, A.C., *The Science of and Practice of Welding*, 9th ed., Cambridge University Press (1989)
Gibson, S. and Smith, A., *Basic Welding*, Macmillan
Gibson, S., *Practical Welding*, Macmillan
Gibson, S.W., *Advanced Welding*, Macmillan (1997)
Gourd, L.M., *Principles of Welding Technology* (1980)
Manko, H.H., *Solders and Soldering,* 2nd ed. (1979)

Automatic control
Coffin, M.J., *Direct Digital Control for Building HVAC Systems*, 2nd ed., Kluwer Academic Publishers (1998)
Fisk, D.J., *Thermal Control of Buildings*, Applied Science Publishers, London (1981)
Levenhagen, J.J., *HVAC Control System Design Diagrams*, McGraw-Hill (1999)
Underwood, C.P., *HVAC Control Systems*, E. & F. N. Spon (1999)

17 Standards

British Standards

The following list of British Standards relevant to heating ventilating and air conditioning is based on information available in March 2000. For the latest details reference should be made to the current BSI Catalogue which is published annually.

10: 1962 Flanges and bolting for pipes, valves and fittings (obsolescent)

Flanges in grey cast iron, copper alloy and cast or wrought steel for $-328°F$ ($-200°C$) to $975°F$ ($524°C$) and up to 2800 lb/in^2. Materials and dimensions of flanges, bolts and nuts. Ten tables cover plain, boss, integrally cast or forged, and welding neck types.

21: 1985 Pipe threads for tubes and fittings where pressure-tight joints are made on the threads (metric dimensions)

Range of threads from 1/16 to 6, together with thread forms, dimensions, tolerances, and designations. Requirements for jointing threads for taper external threads, for assembly with either taper or parallel internal threads and for longscrews specified in BS 1387.

41: 1973 (1998) Cast iron spigot and socket flue or smoke pipes and fittings

Material, dimensions and tolerances of pipes, bends and offsets up to 300 mm nominal bore and nominal weight of pipes.

143 & 1256: 1986 Malleable cast iron and cast copper alloy threaded pipe fittings

Requirements for design and performance of pipe fittings for design to BS 143 having taper external and internal threads and BS 1256 having taper external and parallel internal threads.

417: -- Galvanised low carbon steel cisterns, cistern lids, tanks and cylinders

417: Part 1: 1964 Imperial units (obsolescent)

Cold and hot water storage vessels for domestic purposes.
 Cisterns: 20 sizes from 4 to 740 gallons capacity, in two grades.
 Tanks: 5 sizes from 31 to 34 gallons capacity, in two grades.
 Cylinders: 10 sizes from 16 to 97 gallons capacity, in two grades.

417: Part 2: 1987 Metric units

Capacities from 18 l to 3364 l for cisterns, 95 to 115 l for tanks, 73 to 441 l for cylinders.

499: -- Welding terms and symbols

499: Part 1: 1991 Glossary for welding, brazing and thermal cutting

Gives terms common to more than one process, terms relating to welding with pressure, fusion welding, brazing, testing, weld imperfections and thermal cutting.

499: Part 1: 1992 Supplement

Definitions for electric welding equipment.

499: Part 2c: 1980 Welding symbols

Provides, in chart form, the type, position and method of representation of welding symbols and examples of their use.

567: 1973 (1989) Asbestos - cement flue pipes and fittings, light quality

Diameters 50 to 150 mm, for use with gas-fired appliances up to 45 kW.

599: 1966 Methods of testing pumps

Testing performance and efficiency of pumps for fluids which behave as homogeneous liquids.

699: 1984 (1990) Copper direct cylinders for domestic purposes

Requirements for copper direct cylinders, with capacities between 74 and 450 litres, for storage of hot water. Covers 4 grades and 16 sizes and also factory-applied insulation and protector rods.

715: 1993 Metal flue pipes, fittings, terminals and accessories for gas-fired appliances with a rated input not exceeding 60 kW

749: 1969 Underfeed stokers

Stokers rated up to 550 kg of coal per hour for all furnaces except metallurgical or other high temperature; requirements, installation, maintenance.

759: -- Valves, gauges and other safety fittings for application to boilers and to piping installations for and in connection with boilers

759: Part 1: 1984 Valves, mountings and fittings for boilers

Requirements for safety fittings excluding safety valves for boiler installations where steam pressure exceeds 1 bar gauge or, in the case of hot water boilers, the rating is 44 kW and above.

779: 1989 Cast iron boilers for central heating and indirect water supply (rated output 44 kW and above)

Design and construction including materials, workmanship, inspection, testing and marking of boilers for use with solid, gaseous and liquid fuels.

799: -- Oil burning equipment

799: Part 2: 1991 Vaporising burners

Requirements for oil vaporising burners and associated equipment for boilers, heaters, furnaces, ovens and similar static flued plant such as free standing space-heating appliances, for single family dwellings.

799: Part 3: 1981 Automatic and semi-automatic atomising burners up to 36 litres per hour

Requirements for materials for all component parts and such parts of component design and plant layout as are fundamental to the proper functioning of such equipment.

799: Part 4: 1991 Atomising burners (other than monobloc type) together with associated equipment for single burner and multi-burner installations

For land and marine purposes. Suitable for liquid fuels to BS 2869 and BS 1469.

799: Part 5: 1989 Oil storage tanks

Requirements for carbon steel tanks for the storage of liquid fuel used in conjunction with oil-burning equipment. Includes integral tanks which form part of a complete oil-fired unit, service tanks, and storage tanks with a maximum height of 10 m and capacities up to 150 000 l.

799: Part 7: 1988 Dimensions of atomising oil-burner pumps with rotating shaft and external drive

Fixes the dimensions for connectors and certain dimensional characteristics of pumps.

799: Part 8: 1988 Connecting dimensions between atomising oil burners and heat generators

Applicable to atomising oil burners up to 150 kW capacity.

835: 1973 (1989) Asbestos cement flue pipes and fittings, heavy quality

Diameters from 75 to 600 mm for use with solid fuel and oil-burning appliances of output rating not exceeding 45 kW, for gas-fired appliances and for incinerators not exceeding 0.09 m³ capacity.

845: -- Methods of assessing thermal performance of boilers for steam hot water and high temperature heat transfer fluids

845: Part 1: 1987 Concise procedure

A concise but complete test method, at minimum cost, for assessing the thermal performance of boilers, generally at output greater than 44 kW, which are thermodynamically simple and fired by solid, liquid or gaseous fuels.

845: Part 2: 1987 Comprehensive procedure

A comprehensive test method for assessing the thermal performance of any boiler, generally of output greater than 44 kW, including those with multiple thermal flows to and from the boiler, and fired by solid, liquid or gaseous fuels.

848: --   Fans for special purposes

848: Part 1: 1997 Performance testing using standardised airways

848: Part 2: 1985 Methods of noise testing

Determination of the acoustic performance of fans operating against differences of pressure. Four methods are described: in-duct, reverberant field, free field and semi-reverberant.

848: Part 4: 1997 Dimensions

848: Part 5: 1986 Guide for mechanical and electrical safety

Fans connected to single-phase a.c., three-phase a.c. and d.c. supplies up to 660 V. Identifies the circumstances in which safety measures should be taken and gives information on how safety hazards can be reduced or eliminated.

848: Part 6: 1989 Method measurement of fan vibration

853: --   Vessels for use in heating systems

853: Part 1: 1996 Calorifiers and storage vessels for central heating and hot water supply

Strength and method of construction.

853: Part 2: 1996 Tubular heat exchangers and storage vessels for building and industrial services

855: 1976 Welded steel boilers for central heating and indirect hot water supply (rated output 44 kW to 3 MW)

Requirements for design and construction of boilers for use with solid, gaseous and liquid fuels.

1010: --   Draw off valves and stop valves for water services (screwdown pattern)

1010: Part 2: 1973 Draw off taps and above ground stop valves

Dimensions and test requirements for screwdown pattern draw off taps and above ground stop valves 1/4 in to 2 in nominal sizes. Material, design, dimensions of components and union ends.

1181: 1989 Clay flue linings and flue terminals

For use with certain domestic appliances, including gas-burning installations, and for ventilation. Dimensions, performance characteristics, sampling, testing, inspection and marking.

1212: --   Float operated valves (including floats)

1212: Part 1: 1990 Piston type

Seven sizes from 3/8 in to 2 in. Materials, quality, workmanship, dimensions and performance requirements.

1212: Part 2: 1990 Diaphragm type (copper alloy body)

Workmanship, dimensions and performance requirements for nominal sizes 3/8 and 1/2.

1212: Part 3: 1990 Diaphragm type (plastics body) for cold water services

Operational requirements.

1339: 1965 (1981) Definitions, formulae and constants relating to the humidity of air

Includes tables of saturation vapour pressure and bibliography.

1387: 1985 (1990) Screwed and socketed steel tubes and tubulars and plain end steel tubes suitable for welding or for screwing to BS 21 pipe threads

Applicable to tubes of nominal sizes DN 8 to DN 150 in light, medium and heavy thicknesses.

1394: – – Stationary circulation pumps for heating and hot water service systems

1394: Part 2: 1987 Physical and performance requirements

Physical and performance requirements for pumps with a rated input not exceeding 300 W.

1415: – – Mixing valves

1415: Part 1: 1976 Non-thermostatic, non-compensatory mixing valves

Performance requirements, materials and methods of specification of 1/2 and 3/4 nominal size valves.

1415: Part 2: 1986 Thermostatic mixing valves

Materials, designs, construction and performance requirements, with method of specifying size, for thermostatic mixing valves suitable for use with inlet supply pressures up to 6 bar and inlet water temperature between 10°C and 72°C.

1564: 1975 (1983) Pressed steel sectional rectangular tanks

Working under a pressure not exceeding the static head corresponding to the depth of the tank, built up from pressed steel plates 1220 mm square. The sectional dimensions are interchangeable with the imperial dimensions of the previous standard.

1565: – – Galvanised mild steel indirect cylinders

1565: Part 1: 1949 Imperial units (obsolescent)

1565: Part 2: 1973 Metric units (obsolescent)

1566: -- **Copper indirect cylinders for domestic purposes**

1566: Part 1: 1984 (1990) Double feed indirect cylinders

Requirements for cylinders with capacities between 72 and 440 l for hot water storage. Covers 4 grades and 16 sizes and also includes factory-applied insulation and protector rods.

1566 : Part 2: 1984 Single-feed indirect cylinders

Requirement for 3 grades for cylinders with capacities from 86 to 196 l. Covers 3 grades and 7 sizes. The cylinders are of the type in which the bottom is domed inwards.

1586: 1982 Methods of performance testing and presentation of performance data for refrigerant condensing units

Applies to air and water-cooled condensing units employing single-stage refrigerant compressors including hermetic, semi-hermetic and open types. Describes the method for presentation of performance data for these units including correction factors and part load characteristic where applicable.

1608: 1990 Electrically-driven refrigerant condensing units

Design, construction and testing of units up to a power input of approximately 25 kW.

1710: 1984 (1991) Identification of pipelines and services

Colours for identifying pipes conveying fluids in liquid or gaseous form in land and marine installations.

1740: -- **Wrought steel pipe fittings**

1740: Part 1: 1971 (1990) Metric units

Welded and seamless fittings 6 mm to 150 mm for use with steel tubes to BS 1387, screwed BSP thread to BS 21.

1756: -- **Methods for sampling and analysis of flue gases**

1756: Part 1: 1971 Methods of sampling

1756: Part 2: 1971 Analysis by the Orsat apparatus

Apparatus, reagents, method, sample analysis, calculations, reporting of results.

1756: Part 3: 1971 Analysis by the Haldane apparatus

Apparatus, reagents, method, sample analysis, calculation, reporting of results.

1756: Part 4: 1977 Miscellaneous analyses

Determination of moisture content, sulphuric acid dew point, carbon monoxide, oxides of sulphur and oxides of nitrogen.

1756: Part 5: 1971 Semi-routine analyses

Carbon dioxide, carbon monoxide and total oxides of sulphur. Mainly for combustion performance of domestic gas appliances.

1894: 1992 Design and manufacture of electric boilers of welded construction

Materials, workmanship, inspection, testing, documentation and marking of boilers utilising electrodes or immersion elements to provide hot water or steam. Boilers are cylindrical constructed from carbon or carbon manganese steel by fusion welding.

2051: -- Tube and pipe fittings for engineering purposes

2501: Part 1: 1973 (obsolescent) Copper and copper alloys capillary and compression tube fittings for engineering purposes

Applies to capillary and compression fittings, in sizes from 4 mm to 42 mm. These fittings are intended primarily for use with tubes of the outside diameters given in BS 2871: Part 2.

2740: 1969 (1991) Single smoke alarms and alarm metering devices

Requirements for the construction and operation of instruments designed to give an alarm when smoke emission from a chimney exceeds a chosen Ringelmann shade.

2742: 1969 (1991) Notes on the use of the Ringelmann and miniature smoke charts

Explains the purpose and method of use of these charts for the visual assessment of the darkness of smoke emitted from chimneys.

2742C: 1957 (1991) Ringelmann chart

2742M: 1960 (1991) Miniature smoke chart

A chart, printed in shades of grey matt lacquer, which when held at about 5 ft from the observer gives readings of the density values of smoke from chimneys.

2767: 1991 Manually operated copper alloy valves for radiators

Designation, pressure and temperature ratings, materials, design, construction and testing of manual valves. Includes handwheel torque strength test, connections for metric copper tubes, compression type tailpiece connections, plating and drainage facility.

2811: 1969 (1991) Smoke density indicators and recorders

Requirements of construction and operation of instruments designed to measure the optical density of, or percentage obscuration caused by, smoke emitted from chimneys.

2869: 1998 Fuel oils for agricultural, domestic and industrial engines and boilers

2879: 1980 (1988) Draining taps (screw down pattern)

Specifies 1/2 and 3/4 nominal size copper alloy bodied taps for draining down hot and cold water installations and heating systems.

3048: 1958 Code for the continuous sampling and automatic analysis of flue gases, indicators and recorders

Automatic instruments for direct indication or record of composition of flue gases from industrial plant. Thermal conductivity instruments,

instruments depending on chemical absorption and chemical reaction, viscosity and density instruments, oxygen meters, infra-red absorption instruments. Determination of dew point.

3198: 1981   Copper hot water storage combination units for domestic purposes

Requirements for direct, double-feed indirect and single-feed indirect types of units having hot water storage capacities between 65 and 180 litres.

3250: --   Methods for the thermal testing of domestic solid fuel burning appliances

3250: Part 1: (1993)   Flue loss method

3250: Part 2: 1961 (1988)   Hood method

Describes a method in which the convection warm air from the appliance is collected by means of a hood and measured directly.

3300: 1974   Kerosine (paraffin) unflued space heaters, cooking and boiling appliances for domestic use

Construction, safety, performance, marking and methods of test.

3377: 1985   Boilers for use with domestic solid mineral fuel appliances

Materials, construction and pressure testing of boilers of normal and high output for use with domestic solid mineral fuel appliances.

3416: 1991   Bitumen-based coatings for cold applications, suitable for use in contact with potable water

Two types each with three classes of coatings, all of which give films that comply with the national requirements for contact with potable water.

3505: 1986   Unplasticised polyvinyl chloride (PVC-U) pressure pipes for cold potable water

Pipes up to and including nominal size 24 for use at pressures up to 15 bar and 20°C, such that pipes which conform to the standard will be acceptable to UK water undertakings.

3974: --   Pipe supports

3974: Part 1: 1974   Pipe hangers, slider and roller type supports

Requirements for the design and manufacture of components for the hanger, slide and roller type supports for uninsulated and insulated steel and cast iron pipes of nominal sizes 15 mm to 160 mm within the temperature range −20°C + 470°C.

3974: Part 2: 1978   Pipe clamps, cages, cantilevers and attachments to beams

Applies to pipes of nominal sizes 100 mm to 600 mm.

4127: 1993 Light gauge stainless steel tubes, primarily for water applications

4213: 1991 Cold water storage and combined feed and expansion cistern (polyolefin or olefin copolymer) up to 500 L capacity used for domestic purposes

Requirements for materials and physical properties for cisterns for use in the storage of water.

4256: -- Oil burning air heaters

4256: Part 2: 1972 (1980) Fixed, flued, fan-assisted heaters

Construction, operation, performance and safety requirements for heaters, designed for use with distillate coals such as kerosene, gas oil and domestic fuel oil.

4433: -- Domestic solid mineral fuel fired boilers with rated outputs up to 45 kW

4433: Part 1: 1994 Boilers with undergrate ash removal

4433: Part 2: 1994 Gravity feed boilers designed to burn small anthracite

4485: -- Water cooling towers

4485: Part 2: 1988 Methods of performance testing

Determination of the performance of industrial mechanical draught and natural draught towers.

4504: -- Circular flanges for pipes, valves and fittings

4504: Section 3.1: 1989 Steel flanges

Types of circular steel flanges from PN 2.5 to PN 40 and in sizes up to DN 4000. Facings, dimensions, tolerances, threading, bolt sizes, marking and materials for bolting and flange materials with associated pressure/temperature ratings.

4504: Section 3.2: 1989 Cast iron flanges

Flanges in grey, malleable and ductile cast iron from PN 2.5 to PN 40 and in sizes up to DN 4000. Facings, dimensions, tolerances, threading, bolt sizes, marking and materials for bolting and flange materials with associated pressure/temperature ratings.

4504: Section 3.3: 1989 Copper alloy and composite flanges

Types of flanges from PN 6 to PN 40 and in sizes up to DN 1800. Facings, dimensions, tolerances, bolt sizes, marking and materials for bolting and flange materials with associated pressure/temperature ratings.

4508: -- Thermally insulated underground pipe lines

4508: Part 1: 1986 Steel-cased systems with air gap

Requirements of design, materials, construction, installation, testing and fault monitoring for steel-cased systems for temperatures exceeding 50°C.

4508: Part 4: 1977 Specific testing and inspection requirements for cased systems without air gap

Testing, inspection and certification of pipe-in-pipe distribution systems with an insulated service or product pipe enclosed in a pressure tight casing.

4543: -- Factory-made insulated chimneys

4543: Part 1: 1990 Methods of test

Methods of test for circular cross sectional metal chimneys supplied in component form needing no site fabrication. Intended for internal use.

4543: Part 2: 1990 Chimneys with flue linings for use with solid fuel fixed appliances

Circular cross sectional chimneys needing no site fabrication for internal use.

4543: part 3: 1990 Chimneys with stainless steel flue lining for use with oil fired appliances

Requirements for chimneys with stainless steel internal and metal external surfaces intended for use with oil fired appliances.

4814: 1990 Expansion vessels using an internal diaphragm for sealed hot water heating system

Requirements for manufacture and testing of carbon steel vessels up to 1000 L: capacity, up to 1000 mm diameter and for use in systems operating up to 6 bar.

4856: -- Methods for testing and rating fan coil units, unit heaters and unit coolers

4856: Part 1: 1972 (1983) Thermal and volumetric performance for heating duties, without additional ductwork

Methods of carrying out thermal and volumetric tests on forced convection units containing fluid to air heat exchangers and incorporating their own fans. The units are for heating applications and the tests are to be carried out on units in essentially clean conditions.

4856: Part 2: 1975 (1983) Thermal and volumetric performance for cooling duties, without additional ductwork

Coolers as used for cooling and dehumidifying under frost-free conditions, the medium used being water or other heat transfer fluid (excluding volatile refrigerants).

4856: Part 3: 1975 (1983) Thermal and volumetric performance for heating and cooling duties, with additional ductwork

Units for use with additional ducting containing fluid to air heat exchangers and incorporating their own electrically-powered fan system. For heating and cooling application, the latter with or without dehumidification under frost-free conditions.

4856: Part 4: 1997 Determination of sound power levels for fan coil units, unit heaters and unit coolers using reverberating rooms

4857: -- Methods for testing and rating terminal reheat units for air distribution systems

4857: Part 1: 1972 (1983) Thermal and aerodynamic performance

Terminal reheat units with or without flow rate controllers.

4857: Part 2: 1978 (1985) Acoustic testing and rating

Methods of testing and rating for static terminal attenuation, sound generation, upstream and downstream of the unit, radiation of sound from the casing.

4876: 1984 Performance requirements for domestic flued oil burning appliances

Performance requirements and methods of testing for flued oil burning appliances (e.g. boilers and air heaters) up to and including 44 kW capacity, used for hot water supply and space heating.

4954: -- Methods for testing and rating induction units for air distribution systems

4954: Part 1: 1973 (1987) Thermal and aerodynamic performance

Methods of test for induction units with water coils for heating and/ or sensible cooling duties.

4954: Part 2: 1978 (1987) Acoustic testing and rating

Methods of acoustic testing and rating of induction units for sound power emission and terminal attenuation.

4979: 1986 Methods for aerodynamic testing of constant and variable dual or single duct boxes, single duct units and induction boxes for air distribution systems

Methods of test for casing leakage, valve and damper leakage, flow rate control, temperature mixing, induction flow rate and pressure requirements.

5041: -- Fire hydrant systems equipment

5041: Part 1: 1987 Landing valves for wet risers

Material, design and performance requirements for copper alloy globe and diaphragm valves for wet rising mains. Covers high and low pressure types.

5041: Part 2: 1987 Landing valves for dry risers

Material and design requirements for copper alloy gate valves for dry rising mains.

5041: Part 3: 1975 (1987) Inlet breechings for dry riser inlets
Requirements for 2 and 4-way inlet breechings on a dry rising water main for fire fighters.

5041: Part 4: 1975 (1987) Boxes for landing valves

Dimensions to provide clearances and ensure that valves are easily accessible. Constructional details, requirements for hingeing, glazing, marking, locking of doors.

5041: Part 5: 1974 (1987) Boxes for foam inlets and dry riser inlets

Standard sizes according to the number of inlets for foam or to the size of the riser. Choice and thickness of material. Dimensions of glass in the door frame and marking thereon. May also be used for other purposes, e.g. fuel oil inlets and drencher systems.

5114: 1975 (1981) Performance requirements for joints and compression fittings for use with polyethylene pipes

Resistance to hydraulic pressure, external pressure and pull-out of assembled joints and effect on water and opacity.

5141: -- Air heating and cooling coils

5141: Part 1: 1975 (1983) Method of testing for rating cooling coils

Duct-mounted cooling coil rating test with chilled water as the cooling medium within specified ranges of variables for inlet air and water temperatures and for water flow and air velocity.

5141: Part 2: 1977 (1983) Method of testing for rating heating coils

Rating test for duct-mounted air heating coils with hot water or dry saturated steam as the heating medium.

5258: -- Safety of domestic gas appliances

5258: Part 1: 1986 Central heating boilers and circulators

Safety requirements and associated test methods for natural draught and fan-powered boilers of rated heat input up to 60 kW and for circulators of rated heat input not exceeding 8 kW for circulators.

5258: Part 5: 1989 Gas fires

Safety requirements and associated test methods for open-flued radiant and radiant convector gas fires.

5258: Part 7: 1977 Storage water heaters

Safety requirements and associated test methods for domestic appliances having an input rating not exceeding 20 kW.

5258: Part 9: 1989 Combined appliances: fanned circulation ducted air heaters/circulators

Safety requirements and associated methods of test for heaters either combined with, or designed to be fitted with, circulators. For rated heat inputs not exceeding 60 kW and 8 kW for circulators.

5258: Part 13: 1986 Convector heaters

Requirements and associated test methods for flued natural draught and fan-powered heaters of input rating not exceeding 25 kW.

5410: -- Code of practice for oil firing

5410: Part 1: 1977 Installations up to 44 kW output for space heating and hot water supply

5410: Part 2: 1978 Installation of 44 kW and above output for space heating, hot water and steam supply

Deals with provision of oil-burning systems for boiler and warm air heater plants and associated oil tanks.

5422: 1990 Thermal insulating material on pipes, ductwork and equipment (in the temperature range $-40°C$ to $+700°C$)

Insulation of surfaces of process plant, vessels, tanks, ducts, pipelines, boilers, ancillary plant. Domestic, commercial and industrial applications for heating fluids, steam and refrigeration and air conditioning.

5433: 1976 Underground stop valves for water services

Copper alloy screwdown stop valves, nominal sizes 1/2 to 2.

5440: -- Installation of flues and ventilation for gas appliances of rated input not exceeding 60 kW.

5440: Part 1: 1990 Installation of flues

Complete flue equipment from the appliance connection to the discharge to outside air.

5440: Part 2: 1989 Installation of ventilation for gas appliances

Air supply requirements for domestic and commercial gas appliances installed in rooms and other internal spaces and in purpose designed compartments.

5449: 1990 Forced circulation hot water central heating systems for domestic premises

General planning, design considerations, materials, appliances and components, installation and commissioning. Includes small bore and microbore systems, open and sealed systems.

5588: -- Fire precautions in the design and construction of buildings

5588: Part 9: 1989 Code of practice for ventilation and air conditioning duct work

Recommendations to limit the potential for the spread of fire and its by products.

5615: 1985 Insulating jackets for domestic hot water storage cylinders

Performance in respect of maximum permitted heat loss, materials, design and marking of jackets for cylinders to BS 699 and BS 1566.

5720: 1979 Code of practice for mechanical ventilation and air conditioning in buildings

General design, planning, installation, testing and maintenance of mechanical ventilating and air conditioning systems. Covers general matters, fundamental requirements, design considerations, types and selection of equipment, installation, inspection, commissioning and testing, operation and maintenance, overseas projects.

5864: 1989 Installation in domestic premises of gas-fired ducted-air heaters of rated input not exceeding 60 kW

Selection, installation, inspection and commissioning. Includes commentary and recommendations.

5885: -- Automatic gas burners

5885: Part 1: 1988 Burners with input rating 60 kW and above

Safety aspects for burners employing forced or mechanically-induced draught, packaged or non-packaged types. Covers single burners and dual fuel burners when operating only on gas.

5885: Part 2: 1987 Packaged burners with input rating 7.5 kW up to 60 kW

Requirements for small packaged burners employing forced and induced draught. Covers single burners and dual fuel burners when operating only on gas.

5978: -- Safety and performance of gas-fired hot water boilers (60 kW to 2 MW input)

5978: Part 1: 1989 General requirements

Performance, safety and methods of test for burners operating at internal pressures up to 4.5 bar.

5978: Part 2: 1989 Additional requirements for boilers with atmospheric burners

5978: Part 3: 1989 Additional requirements for boilers with forced or induced draught burners

5990: 1989 Direct gas fired forced convection air heaters with rated inputs up to 2 MW for industrial and commercial space heating

Safety and performance requirements.

5991: 1989 Indirect gas fired forced convection air heaters with rated input up to 2 MW for industrial and commercial space heating

Safety and performance requirements and methods of test for permanently installed open flued space heating appliances for industrial and commercial applications.

6144: 1990 Expansion vessels using an internal diaphragm for unvented hot water supply systems

Requirements for manufacture and testing of steel vessels for use in systems operating with maximum pressure up to 10 bar.

6230: 1991 Installation of gas-fired forced convection air heaters for commercial and industrial space heating of rated input exceeding 60 kW

Requirements for the selection and installation of direct and indirect fired air heaters with or without ducting and with or without recirculation of heated air.

6283: -- Safety and control devices for use in hot water systems

6283: Part 1: 1991 Expansion valves for pressures up to and including 10 bar

Design, construction and testing of expansion valves of the automatic reseating type, specifically intended for preventing overpressurisation due to expansion of water in storage water heaters of the unvented type.

6283: Part 2: 1991 Temperature relief valves for pressures from 1 bar to 10 bar

Design, construction and testing of temperature relief valves of the automatic reseating type, specifically intended for use with and protection of storage water heaters of the unvented type.

6283: Part 3: 1991 Combined temperature and pressure relief valves for pressures from 1 bar to 10 bar

Design, construction and testing of combined temperature and pressure relief valves of the automatic reseating type, specifically intended for use with and protection of storage water heaters of the unvented type.

6283: Part 4: 1991 Drop tight pressure reducing valves of nominal sizes up to and including DN50 for pressures up to and including 12 bar

Design, construction and testing of drop tight pressure reducing valves, sometimes known as pressure limiting valves.

6332: -- Thermal performance of domestic gas appliances

6332: Part 1: 1988 Thermal performance of domestic heating boilers and circulators

Thermal efficiency and associated methods of test for boilers and circulators of rated heat input up to and including 60 kW and 8 kW respectively.

6332: Part 4: 1983 Thermal performance of independent convector heaters

Requirements and associated methods of test for heaters operating under natural draught.

6332: Part 6: 1990 Thermal performance of combined appliances: fanned circulation ducted air heater/circulator

Thermal efficiency requirements and associated methods of test for appliances of rated heat input not exceeding 60 kW with circulators not exceeding 8 kW.

6583: 1985 Methods for volumetric testing for rating of fan sections in central station air-handling units

Definitions of test unit and test installations, methods of test, presentation of data, extrapolation of data for geometrically similar sections and for those which are geometrically similar except for the fans, guide to the rating of air-handling units.

6644: 1991 Installation of gas fired hot water boilers of rated inputs between 60 kW and 2 MW

Requirements for the installation of single and groups of boilers, selection and siting, open vented and sealed systems, controls and safety, air supply and ventilation, flues and commissioning.

6675: 1986 Servicing valves (copper alloy) for water services

Three patterns of servicing valves for isolation of water supplies to individual sanitary appliances so that those appliances can be maintained or serviced.

6700: 1987 Design, installation, testing and maintenance of services supplying water for domestic use within buildings

System of pipes, fittings and connected appliances installed to supply any building with hot and cold water for general purposes.

6759: -- Safety valves

6759: Part 1: 1984 Safety valves for steam and hot water

Requirements for safety valves for boilers and associated pipework for steam pressures exceeding 1 bar and hot water boilers of rating 44 kW and above.

6798: 1987 Installation of gas-fired hot water boilers, of rated input not exceeding 60 kW

Selection, installation, inspection and commissioning of gas-fired central heating installations for domestic or commercial premises by circulation of heated water.

6880: -- Code of practice for low temperature hot water heating systems of output greater than 45 kW

6880: Part 1: 1988 Fundamental design considerations

Requirements which need to be taken into account in the design of open vented or sealed systems.

6880: Part 2: 1988 Selection of equipment

Types of low temperature hot water heating equipment in common use and the selection of such equipment.

6880: Part 3: 1988 Installation, commissioning and maintenance

Recommendations for installation, commissioning, operation and maintenance of open vented or sealed systems.

6896: 1991 Installation of gas-fired overhead radiant heaters for industrial and commercial heating

Installation, inspection and commissioning of heaters for other than domestic premises.

7074: -- Application, selection and installation of expansion vessels and ancillary equipment for sealed water systems

7074: Part 1: 1989 Code of practice for domestic heating and hot water supply

7074: Part 2: 1989 Code of practice for low and medium temperature hot water heating systems

Vessels and systems for heating larger premises, commercial and industrial.

7074: Part 3: 1989 Code of practice for chilled and condenser systems

Vessels and systems for air conditioning of commercial and industrial premises.

7186: 1989 Non domestic gas fired overhead radiant tube heaters

Safety and construction and associated methods of test.

7206: 1990 Unvented hot water storage units and packages

Requirements for units and packages heated directly or indirectly. Cylinders with capacities from 15 L to 500 L having minimum opening pressure of 1 bar and minimum operating pressure of 3 or 6 bar, fitted with safety devices to prevent water temperature from exceeding 100°C.

7291: -- Thermoplastic pipes and associated fittings for hot and cold water for domestic purposes and heating installations in buildings

7291: Part 1: 1990 General requirements

General requirements and application classes for pipes up to 67 mm in alternative metric sizes for plastics or copper. Performance requirements for associated fittings.

7291: Part 2: 1990 Polybutylene (PB) pipes and associated fittings

Pipes 10 mm to 53 mm outside diameter.

7291: Part 3: 1990 Crosslinked polyethylene (PE-X) pipes and associated fittings and solvent cement

Pipes 12 mm to 35 mm outside diameter. Includes cemented joints.

7291: Part 4: 1990 Chlorinated polyvinylchloride (PVC) pipes and associated fittings and solvent cement

Pipes 12 mm to 63 mm outside diameter. Includes cemented joints.

7350: 1990 Double regulating globe valves and flow measurement devices for heating and chilled water

Pressure and temperature rating, materials, performance requirements, testing, marking, installation and operating instructions.

7478: 1991 Guide to selection and use of thermostatic radiator valves

Guidance on selection, application and use of thermostatic radiator valves.

7491: -- Glass reinforced plastics cisterns for cold water storage

7491: Part 1: 1991 One piece cisterns of capacity up to 500 L

7491: Part 2: 1992 One piece cistern of nominal capacity from 600 L to 25 000 L

7491: Part 3: 1994 Sectional tanks

7556: 1992 Thermoplastic radiator valves. Specification for dimensions and details on connection

7593: 1992 Code of practice for treatment of water in domestic hot water central heating systems

Guidance for the preparation of wet central heating systems prior to use, and for application of inhibitors.

8313: 1989 Code of practice for accommodation of building services in ducts

Recommendation for design, construction, installation and maintenance.

European Standards

BS EN 215: -- Thermostatic radiator valves

BS EN 215-1: 1991 Requirements and test methods

Definitions, mechanical properties, operating characteristics and test methods.

BS EN 253: -- Preinsulated bonded pipe systems for underground hot water networks. Pipe assembly of steel service pipes, polyurethane thermal insulation and outer casing of high density polyethylene.

Requirements and test methods for straight lengths of prefabricated thermally insulated pipe-in-pipe assemblies.

BS EN 297: 1994 Gas fired central heating boilers fitted with atmospheric burners of nominal heat input not exceeding 70 kW

BS EN 303: -- Heating boilers. Heating boilers with forced draught burners

BS EN 303-1: 1992 Terminology, general requirements, testing and marking

BS EN 303-2: 1992 Special requirements for boilers with atomising oil burners

BS EN 304: 1992 Heating boilers, Test code for heating boilers for atomising oil burners

BS EN 378: -- Refrigerating systems and heat pumps. Safety and environmental requirements

BS EN 378-1: 1995 Basic requirements

BS EN 442: -- Radiators and convectors

BS EN 442-1: 1996 Technical specifications and requirements

BS EN 442-1: 1997 Test methods and rating

BS EN 442-3: 1997 Evaluation of conformity

BS EN 448: 1995 Preinsulated bonded pipe systems for underground hot water networks. Fittings assemblies of steel service pipes, polyurethane thermal insulation and outer casing of polyethylene

BS EN 488: 1995 Preinsulated bonded pipe systems for underground hot water networks. Steel valve assembly of steel service pipes, polyurethane thermal insulation and outer casing of polyethylene

BS EN 489: 1995 Preinsulated bonded pipe systems for underground hot water networks. Joint assembly for steel service pipes, polyurethane thermal insulation and outer casing of polyethylene

BS EN 525: 1998 Non domestic direct gas fired convection air heaters for space heating not exceeding a net heat input of 300 kW

BS EN 621: 1998 Non domestic gas fired forced convection air heaters for space heating not exceeding a net heat input of 300 kW, without a fan to assist transportation of combustion air and/or combustion products

BS EN 625: 1996 Gas fired central heating boilers. Specific requirements for the domestic hot water operation of combination boilers of nominal heat input not exceeding 70 kW

BS EN 677: 1998 Gas fired central heating boilers. Specific requirements for condensing boilers with a nominal heat input not exceeding 70 kW

BS EN 778: 1998 Domestic gas fired forced convection air heaters for space heating not exceeding a net heat input of 70 kW, without a fan to assist transportation of combustion air and/or combustion products

BS EN 779: 1983 Particulate air filters for general ventilation. Requirements, testing, marking

BS EN 1020: 1998 Non domestic gas fired forced convection air heaters for space heating not exceeding a net heat input of 300 kW, incorporating a fan to assist transportation of combustion air and/or combustion products

BS EN 1057: 1996 Copper and copper alloys. Seamless round copper tubes for water and gas in sanitary and heating applications

BS EN 1196: 1998 Domestic and non-domestic gas fired air heaters. Supplementary requirements for condensing air heaters

BS EN 1254: -- Copper and copper alloys. Plumbing fittings

BS EN 1254-1: 1998 Fittings with ends for capillary soldering or capillary brazing to copper tubes

BS EN 1254-2: 1998   Fittings with compression ends for use with copper tubes

BS EN 1254-3: 1998   Fittings with compression ends for use with plastics pipes

BS EN 1254-4: 1998   Fittings combining other end connections with capillary or compression ends

BS EN 1254-5: 1998   Fittings with short ends for capillary brazing to copper tubes

BS EN 1264: --   Floor heating. Systems and components

BS EN 1264-1: 1998   Definitions and symbols

BS EN 1264-2: 1998   Determination of the thermal output

BS EN 1264-3: 1998   Dimensioning

BS EN 1886: 1998   Ventilation for buildings. Air handling units. Mechanical performance

International Standards

BS ISO 4065: 1996   Thermoplastics pipes. Universal wall thickness table

BS EN ISO 5167: --   Measurement of fluid flow by means of pressure differential devices

BS EN ISO 5167-1: 1997   Orifice plates, nozzles and Venturi tubes inserted in circular cross section tubes running full

BS ISO 6243: 1997   Climatic data for building design. Proposed systems of symbols

BS EN ISO 6708: 1996   Pipework components. Definition and selection of DN (nominal size)

BS EN ISO 6946: 1997   Building components and building elements. Thermal resistance and thermal transmittance. Calculation method

BS EN ISO 9251: 1996   Thermal insulation. Heat transfer. Conditions and properties of materials. Vocabulary

BS EN ISO 9288: 1996   Thermal insulation. Heat transfer by radiation. Physical quantities and definitions

BS EN ISO 9300: 1995   Measurement of gas flow by means of critical flow Venturi nozzles

BS EN ISO 10211: --   Thermal bridges in building construction. Heat flows and surface temperatures

BS EN ISO 10211-1: 1996   General calculation methods

BS ISO 11922: -- Thermoplastics pipes for the conveyance of fluids. Dimensions and tolerances

BS ISO 11922-1: 1997 Metric series

BS ISO 11922-2: 1997 Inch-based series

BS EN ISO 13370: 1998 Thermal performance of buildings. Heat transfer via the ground. Calculation methods

BS ISO TR 15377: 1998 Measurement of fluid flow by means of pressure differential devices. Guidelines for specification of nozzles and orifice plates beyond the scope of ISO 5167-1

Describes the geometry and methods of use of various types of orifice plates and nozzles outside the scope of ISO 5167-1.

Index